Schriftenreihe zur Physiolog.. ...

Biochemie der Pflanzen

Herausgeber: Prof. Dr. A. Schaller

Band 12/2019 Fatima Haj Ahmad

Phosphoproteomics Analysis
of the Systemin Signaling Pathway

D 100 (Diss. Universität Hohenheim)

Phosphoproteomics Analysis of the Systemin Signaling Pathway in Tomato

Dissertation for Obtaining the Doctoral Degree

of Natural Sciences (Dr. rer. nat.)

Faculty of Natural Sciences

University of Hohenheim

Institute of Molecular Plant Physiology

Submitted by

Fatima Haj Ahmad

from Amman/Jordan

2019

Dekan: Prof. Dr. Uwe Beifuß
1. berichtende Person: Prof. Dr. Andreas Schaller
2. berichtende Person: Prof. Dr. Waltraud Schulze
Eingereicht am: 10. 05. 2019
Mündliche Prüfung: 28. 06. 2019

Die vorliegende Arbeit wurde am 12. 06. 2019 von der Fakultät
Naturwissenschaften der Universität Hohenheim als „Dissertation zur Erlangung
des Doktorgrades der Naturwissenschaften" angenommen.

Printed and published with the support of the German Academic Exchange
Service (DAAD)

Bibliografische Information der Deutschen Nationalbibliothek

Die Deutsche Nationalbibliothek verzeichnet diese Publikation in der
Deutschen Nationalbibliografie; detaillierte bibliographische Daten sind im Internet
über http://dnb.d-nb.de abrufbar.
1. Aufl. - Göttingen: Cuvillier, 2019
    Zugl.: Hohenheim, Univ., Diss., 2019

© CUVILLIER VERLAG, Göttingen 2019
    Nonnenstieg 8, 37075 Göttingen
    Telefon: 0551-54724-0
    Telefax: 0551-54724-21
    www.cuvillier.de

    ISBN 978-3-7369-7046-5
    eISBN 978-3-7369-6046-6

Table of Contents

Tables

Figures

Abbreviation

ABA	Abscisic Acid
ABI2	ABA-Insensitive2
ACS	1-Aminocyclopropane-L-Carboxylate Synthase
ACX	Acyl-CoA Oxidase
AHA1	*Arabidopsis* H$^+$-ATPase
AOS	Allene Oxide Synthase
APS	Ammonium Persulfate
ARF	Auxin Response Factor 2
At	*Arabidopsis thaliana*
ATP	Adenosine Triphosphate
BAK1	BRI1-Associated Receptor Kinase 1
BAM1	Barely Any Meristem 1
BiFC	Bimolecular Fluorescence Complementation
BIK1	Botrytis-Induced Kinase1
BRI1	Brassinosteroid Insensitive1
BSA	Bovine Serum Albumin
CA	Constitutive Active
CaM	Calmodulin
CaMV	Cauliflower Mosaic Virus
cDNA	Complementary DNA
CDPK	Calcium-Dependent Protein Kinase
cGMP	cyclic GMP
CIPK	CBL-Interacting Protein Kinase
cis-(+)-OPDA	cis-(+)-12-Oxo-Phytodienoic Acid
CNGC17	Cyclic Nucleotide-Gated Channel17
CRISPR/Cas9	Clustered Regularly Interspaced Short Palindromic Repeats/CRISPR-asscoiated protein 9
C-terminus	Carboxyl-terminus
CTR1	Constitutive Triple Response1
DAMPs	Damage-Associated Molecular Patterns
DNA	Deoxyribonucleic Acid
DORN1	Does Not Respond to Nucleotides1

DRMs	Detergent-Resistant Membranes
DTT	Dithiothreitol
eATP	extracellular ATP
EDTA	Ethylenediaminetetraacetic Acid
EFR	Elongation Factor Tu (EF-Tu) Receptor
EGTA	Ethylene Glycol-bis (β-aminoethyl ether)-N,N,N',N'-Tetraacetic Acid
ET	Ethylene
ETI	Effector-Triggered Immunity
EtOH	Ethanol
FACs	Fatty acid–Amino acid Conjugates
FDR	False Discovery Rate
Flg22	Flagellin N-terminal 22-amino acids
FLS2	Flagellin-Sensing 2
GC	Guanylate Cyclase
GHR1	Guard Cell Hydrogen Peroxide Resistant1
HAMPs	Herbivore-Associated Molecular Patterns
HEPES	4-(2-Hydroxyethyl)-1-Piperazineethanesulfonic Acid
13-HPOT	(13-S)-Hydroperoxy Linolenic Acid
HR	Hypersensitive Response
HRP	Horseradish Peroxidase
HT1	High Leaf Temperature1
HypSys	Hydroxyproline-rich Systemins
ID	Intracellular Domain
IMAC	Immobilized Metal Affinity Chromatography
IPTG	Isopropyl β-D-1-Thiogalactopyranoside
JA	Jasmonic Acid
JA-Ile	JA-Isoleucine
JM	Juxtamembrane
KAT	3-Ketoacyl-CoA Thiolase
KEU	SNARE-interacting Protein KEULE
KO	Knock Out
LAP-A	Leucine Aminopeptidase A
LB	Lysogeny Broth
LC-MS/MS	Liquid Chromatography coupled with tandem Mass Spectrometry

LHA1	*Solanum lycopersicum* H^+-ATPase
LOX	Lipoxygenase
LPS	Lipopolysaccharide
LRK10L1.2	Leaf Rust 10 Disease-Resistance Locus Receptor-Like Kinase1.2
LRR-Rlks	Leucine Rich Repeat Receptor-Like Kinase
LRRXIV	Leucine Rich Repeat class XIV
LYK4	Lysin Motif (LysM)-containing RLK4
MAMPs	Microbe-Associated Molecular Patterns
MAPK/MPK	Mitogen-Activated Protein Kinase
MBP	Myelin-Basic Protein
MD	Molecular Dynamics
MeJA	Methyl-Jasmonate
MFP	Multifunctional Proteins
MOAC	Metal Oxide Affinity Chromatography
mRNA	messenger Ribonucleic Acid
MS	Mass Spectrometry
MS-Salts	Murashige and Skoog basal salts
NAD	Nicotinamide Adenine Dinucleotide
NADPH	Nicotinamide Adenine Dinucleotide Phosphate
NAMPs	Nematode-Associated Molecular Patterns
N-terminus	Amino-terminus
OD	Optical Density
OGs	Oligogalacturonides
OPC-8	3-Oxo-2-(20-Pentenyl)-Cyclopentane-1-octanoic acid
OPR3	OPDA-Reductase3
OS	Oral Secretions
PAGE	Polyacrylamide Gel Electrophoresis
PAMPs	Pathogen-Associated Molecular Patterns
ParAMPs	Parasite-Associated Molecular Patterns
PCR	Polymerase Chain Reaction
PD	Phosphatase Domain
PDP1	Pyruvate Dehydrogenase Phosphatase1
PEPR1/2	Plant Elicitor Peptides Receptor1/2
PEPs	Plant Elicitor Peptides

PI	Proteinase Inhibitor
PIPs	PAMP-Induced Peptides
PLA	Phospholipase A
PLL	POLTERGEIST-Like
PM	Plasma Membrane
PORK1	PEPR1/2 Ortholog Receptor-like Kinase1
PP2A	Protein phosphatase 2A
PP2C	Protein phosphatase 2C
PPII	Polyproline II
PPO	Polyphenol Oxidase
PR	Pathogenesis-Related
PRRs	Pattern Recognition Receptors
PSK	Phytosulfokine
PSKR1/2	Phytosulfokine Receptor1/2
PTI	Pattern-Triggered Immunity
qRT-PCR	Quantitative real Time Polymerase Chain Reaction
R	Resistance
RAB	Ras-related protein
RALF	Rapid Alkalinization Factor
RBOH	Respiratory Burst Oxidase Homolog
RLK7	Receptor-Like Kinase7
ROS	Reactive Oxygen Species
SA	Salicylic Acid
SDS	Sodium Dodecyl Sulfate
sgRNA	single guide RNA
Sl	*Solanum lycopersicum*
SlPhyt-1/2	SlPhytaspase-1/2
SYR1/2	Systemin Receptor1/2
Taq	*Thermus aquaticus*
TEMED	Tetramethylethylenediamine
TFA	Trifluoroacetic Acid
TPK1b	Tomato Protein Kinase 1b
UV-B	Ultraviolet B
VAMPs	Vesicle-associated membrane proteins

VEG	Ventral Eversible Gland
VOCs	Volatile Organic Compounds
WAK1	Wall-Associated (Receptor) Kinase1
Wfi1	Whitfly-Induced gp91^{phox}
WIR	Wound-Induced Resistance
WT	Wild Type
X-Gal	5-bromo-4-chloro-3-indoyl-β-D-Galactopyranoside
YAK1	Yeast YAK1-Related Gene 1

Summary

One of the key players involved in herbivore and wound defense responses in tomato is Systemin. It was the first signaling peptide identified in plants in 1991, but the proteins and mechanisms involved in Systemin perception and signal transduction are still poorly understood. To address Systemin-induced signaling events, a phosphoproteomic profiling study involving time-course stimulation of *Solanum peruvianum* cell suspension cultures with Systemin and its inactive analog A17 was performed to reconstruct a Systemin-specific kinase/phosphatase signaling network. The time course analysis of Systemin-induced phosphorylation patterns revealed early events at the plasma membrane, such as dephosphorylation of H^+-ATPase, rapid phosphorylation of NADPH-oxidase and Ca^{2+}-ATPase. Later responses involved transient phosphorylation of small GTPases and vesicle trafficking proteins, as well as transcription factors. Based on a correlation analysis of Systemin-specific phosphorylation profiles, substrate candidates for 56 Systemin-specific kinases and 17 phosphatases were predicted including several receptor kinases as well as kinases with downstream signaling functions, such as MAP-kinases. A regulatory circuit for plasma membrane H^+-ATPase was predicted and confirmed by *in vitro* activity assays. In this regulatory model it is proposed that upon Systemin treatment, H^+-ATPase LHA1 is rapidly de-phosphorylated at its C-terminal regulatory residue T955 by phosphatase PLL5, resulting in the alkalization of the growth medium within two minutes of Systemin treatment. Further, it is suggested that the H^+-ATPase LHA1 is re-activated by MAP-Kinase MPK2 later in the Systemin response. MPK2 was identified with increased phosphorylation at its activating TEY-motif at 15 minutes of treatment and the predicted interaction with LHA1 was confirmed by *in vitro* kinase assays.

The Systemin signaling pathway was addressed also by studying the function of Systemin-induced receptor-like kinases (RLKs), which were selected from the phosphoproteomics data set. The relevance of these candidates as well as of the known

SYR1, SYR2 and PORK1 receptors for early and late Systemin-induced signaling events was analyzed *in vivo* in tomato plants and in *S. peruvianum* cell suspension cultures by creating loss-of-function mutants. An essential function for SYR1 for Systemin perception and early Systemin responses (alkalization of the growth medium) was confirmed in cell cultures. PORK1 function was found to contribute to early Systemin signaling events as well, in addition to its described role in the induction of late responses, including the induction of the proteinase inhibitor II wound response marker. Finally, it is proposed that Systemin crosstalk with other phytohormones, namely ABA and PSK, is mediated through LRK10L1.2 and PSKR2 receptors, respectively.

Zusammenfassung

Systemin ist einer der Schlüsselakteure der Wundantwort bei Tomaten. Es war das erste in Pflanzen identifizierte Peptid-Signal im Jahr 1991. Dennoch sind die Proteine und Mechanismen, die an der Systemin-Wahrnehmung und Signaltransduktion beteiligt sind, bislang kaum verstanden. Zur Untersuchung der Systemin-induzierten Signalwege wurde eine Phosphoproteomik-Profilierungsstudie durchgeführt, die eine zeitliche Stimulierung einer Suspensionskultur von S. peruvianum mit Systemin und dessen inaktivem Analogon A17 beinhaltete, was die Rekonstruktion eines Systemin-spezifischen Kinase/Phosphatase-Signalisierungsnetzwerkes erlaubte. Die Zeitverlaufsanalyse Systemin-induzierter Phosphorylierungsmuster zeigte frühe Ereignisse an der Plasmamembran, wie die Dephosphorylierung der H^+-ATPase und eine schnelle Phosphorylierung der NADPH-Oxidase und Ca^{2+}-ATPase. Spätere Antworten umfassten die transiente Phosphorylierung von kleinen GTPasen und Vesikeltransportproteinen sowie Transkriptionsfaktoren. Basierend auf einer Korrelationsanalyse Systemin-spezifischer Phosphorylierungsprofile konnten mögliche Substrate für 56 Systemin-spezifische Kinasen und 17 Phosphatasen vorhergesagt werden, darunter mehrere Rezeptor-Kinasen sowie Kinasen mit nachgeschalteten Signalfunktionen wie MAP-Kinasen. Ein Regelkreis für die Kontrolle der H^+-ATPase der Plasmamembran wurde vorhergesagt und durch in-vitro-Aktivitätstests bestätigt. In diesem Regulierungsmodell wird vorgeschlagen, dass es nach der Systemin-Behandlung zu einer schnellen Dephosphorylierung der H^+-ATPase LHA1 an ihrem C-terminalen regulatorischen Rest T955 durch die Phosphatase PLL5 kommt, was innerhalb von zwei Minuten nach der Systemin-Behandlung zur Alkalisierung des Wachstumsmediums führt. Weiterhin wird vorgeschlagen, dass die H^+-ATPase LHA1 später in der Systemin-Reaktion durch MAP-Kinase MPK2 wieder aktiviert wird. MPK2 wurde 15 Minuten nach Behandlung mit erhöhter Phosphorylierung an seinem aktivierenden TEY-Motiv identifiziert und die Interaktion mit LHA1 wurde mittels in-vitro Kinasetests bestätigt.

Der Systemin-Signalweg wurde weiterhin im Hinblick auf eine Beteiligung Systemin-induzierter, Rezeptor-artigen Kinasen (RLKs) untersucht, die aus dem Phosphoproteomik-Datensatz hervorgegangen waren. Die Bedeutung dieser Rezeptor-Kandidaten sowie der SYR1, SYR2 und PORK1 Rezeptoren für frühe und späte Systemin-induzierte Signalereignisse wurde *in vivo* in Tomatenpflanzen und *S. peruvianum*-Zellsuspensionskulturen mit Hilfe eines Funktionsverlust-Ansatzes getestet. Die Notwendigkeit von SYR1 für die Systemin-Wahrnehmung und frühe Systemin-Antworten (Alkalisierung des Wachstumsmediums) wurde bestätigt. Es konnte auch eine Beteiligung von PORK1 an frühen Systemin-induzierten Signalereignissen nachgewiesen werden, neben der bekannten Bedeutung von PORK1 für die Regulation der später Anworten, derunter die Induktion des Proteinaseinhibitor II Wundreaktion-Markers. Darüber hinaus wird in dieser Arbeit vorgeschlagen, dass der Systemin-Crosstalk mit anderen Phytohormonen, nämlich ABA und PSK, durch die Rezeptoren LRK10L1.2- bzw. PSKR2-Rezeptoren vermittelt wird.

1. Introduction

Plants form the basis of most food chains on the planet. They influence their environment by shaping weather patterns, providing flood protection, purifying water, and providing food. On the other hand, plant survival is influenced by the environment. In order to survive their ever-changing surroundings while being anchored to the ground, plants have evolved a myriad of strategies to deal with many environmental challenges including thermal stress, wounding, oxidative stress, pathogens and herbivore attacks (Rampitsch and Bykova 2012).

The most prevalent herbivores are herbivorous insects. Their interaction with plants dates back several hundred million years, which has given rise to the co-evolutionary hypothesis (Fürstenberg-Hägg et al. 2013). This hypothesis proposes that insect feeding on plants has been a determining factor in increasing species diversity in both herbivores and hosts (Ehrlich and Raven 1964). Insect herbivores have traditionally been divided into specialists (monophagous and oligophagous insects) that feed on one or more host species from the same family, or generalists (polyphagous insects), which feed on several hosts from different plant families (Fürstenberg-Hägg et al. 2013).

The strategies employed by plants to defend themselves against insect herbivores are very diverse. They include morphological and biochemical responses, which might be either constitutively produced or induced upon attack (Howe and Schaller 2008; Mello and Silva-Filho 2002; War et al. 2018).

1.1. Plant Defense Strategies

Plant morphological defense strategies represent the first defense line of physical barriers against pathogens and herbivores. For example, epicuticular wax films and crystals increase slipperiness preventing insects from populating leaf surfaces (Savatin et al. 2014). Some plants have thorns and spines, which act mainly against mammalian herbivores (Howe and Schaller 2008). Leaf toughness is a physical barrier that exerts a challenge for insects (Howe and Schaller 2008). Upon wounding it is reinforced by deposition and accumulation of macromolecules in the cell wall such as lignin, cellulose, suberin, and callose leading to an induced physical resistance in plants (Howe and

1

Schaller 2008; Pastor et al. 2018). Additionally, trichomes hinder small insects from contacting the leaf surface or limit their movement (Mitchell et al. 2016). It is reported that the density of trichomes increase on young developing leaves of some plant species under herbivore attack to protect the plant from possible future attacks by the second herbivore generation (Dalin and Björkman 2003). On the other hand, glandular trichomes can be considered as both morphological and chemical resistance factors. Glandular trichomes produce and store substances, such as Volatile Organic Compounds (VOCs) and secondary metabolites, which can repel insect herbivores or immobilize them on the leaf surface upon destruction (Howe and Jander 2008; Howe and Schaller 2008).

One of the biochemical and physical barriers are resin and latex, which are viscous organic materials stored under internal pressure in ducts or network of canals in plants (Levin 1976). When herbivore feeding destroys these structures, their contents is released to trap or poison attacking insects and seal the wound to protect it from invading pathogens (Levin 1976; LoPresti 2016).

The major chemicals involved in the biochemical defenses are secondary metabolites (Howe and Jander 2008). Beside their importance as medicinal drugs, poisons, flavors, and industrial materials, the primary function of these chemicals is in plant defense, where they act as deterrents for herbivorous insects, as toxins or as antinutritive substances that reduce the nutritional value of the plant material (Fürstenberg-Hägg et al. 2013; Taiz and Zeiger, 2010). They include a huge number of compounds, which are classified into phenolics, terpenoids or alkaloids (Howe and Schaller 2008; Taiz and Zeiger 2010).

Phenolics such as tannins, coumarins, and phenylpropanoids serve as defense compounds by repelling feeding herbivores, by protection against microbial infections, and as building blocks of lignin (Adeboye et al. 2014). Terpenoids are released from plants under herbivore attack and act as antifeedants, repellents, toxins or as modifiers of insect development (Fürstenberg-Hägg et al. 2013). They are major components of resin and VOCs (Taiz and Zeiger 2010; Yazaki et al. 2017). Volatile bursts of terpenoids can act directly as herbivore repellent, or indirectly by attracting predators that kill plant-feeding insects (Sabelis et al. 2007; Unsicker et al. 2009). VOCs emitted by herbivore-damaged plants into the atmosphere interact with undamaged parts of the same plant as well as the neighboring plants, alerting them to their current or future risk of damage by herbivores (Arimura and Pearse 2017). Alkaloids are believed to provide defense against insects, pathogens and mammalian herbivores because of their general toxicity, deterrence capability and metabolic effects (Mithöfer and Boland 2012). Examples of alkaloids are caffeine, nicotine, morphine, strychnine, and cocaine (Mithöfer and Boland 2012).

Another kind of biochemical defense is the extrafloral nectar, which is secreted on plant leaves and shoots to attract predators and parasitoids serving as one of the indirect defense lines of plants (González-Teuber and Heil 2009). Extrafloral nectar secretion is constitutive in some plant species (Heil et al. 2004), and induced by wounding in others (Heil and Ton 2008). Central American *Acacia* species, for example, are obligately inhabited by symbiotic ants that nourish from constitutively secreted extrafloral nectar and serve as an army that defends their host from herbivore attacks (Heil et al. 2004).

Among the induced biochemical defense strategies in plants are inducible defense proteins that reduce the insect's ability to digest the plant causing amino acid deficiencies, which negatively affect their growth and development (Mithöfer and Boland 2012). The plants' defensive protein arsenal includes enzymes such as arginase and threonine deaminase isoforms (Gonzales-Vigil et al. 2011) that degrade dietary amino acids necessary for insect growth (Howe and Jander 2008; Zhu-Salzman et al. 2008). Enzyme inhibitors such as α-amylase and proteinase inhibitors (PIs) can hinder starch and protein digestion respectively by binding tightly to the digestive enzymes such as α-amylase, trypsin and chymotrypsin (Mithöfer and Boland 2012; Taiz and Zeiger 2010). Oxidative enzymes such as polyphenol oxidase (PPO) and lipoxygenase (LOX) covalently modify the dietary proteins through the production of reactive *o*-quinones and lipid peroxides, respectively (Howe and Jander 2008). Some proteins such as ascorbate oxidases are involved in the disturbance of the insects' gut redox state, which may cause proliferation of oxyradicals that damage proteins, lipids, and DNA of the insect (Howe and Jander 2008; Wang and Constabel 2004). Some of the defense proteins are toxic, they target and damage the peritrophic matrix protecting the midgut epithelium of the insect such as lectins, thionins, chitinases, cysteine proteases and leucine aminopeptidases (Bowles 1990; Zhu-Salzman et al. 2008).

1.2. Induced Innate Immunity against Herbivore Attacks

Induction of defense mechanisms in plants depends on the successful recognition of their enemies by a sophisticated innate immune system (Boller and Felix 2009). Danger signals are recognized by surface-localized Pattern Recognition Receptors (PRRs) that activate resistance responses referred to as Pattern-Triggered Immunity (PTI; Gust et al. 2017). Danger signals include molecular patterns associated with microbes (bacteria, viruses, fungi or oomycetes), nematodes, parasites and herbivores that act as elicitors for plant defense. These elicitors are termed Microbe- or Pathogen-Associated Molecular Patterns (MAMPs/PAMPs; Gust et al. 2017), NAMPs (Manosalva et al. 2015), ParAMPs (Hegenauer et al. 2016), and HAMPs (Basu et al. 2018; Mithöfer and Boland 2008), respectively. A key feature of the molecular signatures of these elicitors is that they are not present in the host, and therefore characterized as 'non-self' (Gust et al. 2017). Endogenous elicitors produced as a result of the damage caused by invading organisms are described as Damage-Associated Molecular Patterns (DAMPs; Boller and Felix 2009).

PTI induced by different elicitors perception via their PRRs provides protection against invaders that are unable to subvert the immune system of a given plant (Andolfo and Ercolano 2015). Host-adapted plant predators such as generalist herbivores, however, have evolved effector proteins as weapons, which target this first defense line at different stages to suppress active immunity, thereby increasing host susceptibility and enabling them to colonize their host (Deslandes and Rivas 2012; Kushalappa et al. 2016). Plants, in consequence, have evolved a second line of resistance called Effector-Triggered Immunity (ETI), which is based on the highly specific direct or indirect interaction of pathogen effectors and the products of plant resistance R genes (Jones and Dangl 2006; Jones et al. 2016).

During feeding on their host plants, insects release a vast array of HAMP elicitors and effectors (Erb et al. 2012). These HAMPs and effectors may arise from insect oral secretions (OS; regurgitant), saliva, ventral eversible gland (VEG) secretions, digestive waste products (e.g., frass) or ovipositional fluids (Basu et al. 2018). They are diverse in structure including: (i) enzymes such as glucose oxidase, ATPase and ß-glucosidase (Eichenseer et al. 1999; Mattiacci et al. 1995), (ii) modified forms of lipids like Fatty acid–Amino acid Conjugates (FACs) such as volicitin (Alborn et al. 1997), (iii) sulfur-containing fatty acids such as caeliferins (Alborn et al. 2007) and (iv) peptides released from digested plant proteins such as inceptins that are proteolytic fragments of the chloroplastic ATP synthase γ-subunit (Schmelz et al. 2006). In addition, the endosymbiotic microbes living in guts of herbivorous insects can induce and manipulate plant defense responses with their released MAMPs and effectors (Acevedo et al. 2015; Chung et al. 2013). Very little is known about PRRs or R proteins that recognize HAMPs or herbivore effectors in plants, respectively (Acevedo et al. 2015; Basu et al. 2018).

Although a putative volicitin receptor in maize has been reported (Truitt et al. 2004), its identity is still unclear (Acevedo et al. 2015; Basu et al. 2018).

Most DAMPs are plant cell components that are released passively upon mechanical damage caused by herbivore-feeding, or by hydrolytic enzymes that are released upon herbivory, microbial or fungal infections, and therefore termed as 'altered self' (Gust et al. 2017). They include the oligomeric fragments of plant cell-wall pectin, termed oligogalacturonides (OGs), which are recognized by WALL-ASSOCIATED (RECEPTOR) KINASE1 (WAK1) in *Arabidopsis* (Ferrari et al. 2013; Kohorn 2016), cutin-derived ω-hydroxy fatty acid monomers (and their corresponding alcohols) and cellulose-derived cellobiose and likely higher-order cellodextrins (Boller and Felix 2009; Choi et al. 2014; Gust et al. 2017). In addition, intracellular molecules such as ATP and NAD$^+$, which are released into the apoplastic space upon cellular damage can be recognized as DAMPs (Gust et al. 2017). In *Arabidopsis*, the extracellular ATP (eATP) is sensed by the L-type lectin receptor kinase Does Not Respond To Nucleotides1 (DORN1) and is able to induce plant defense (Choi et al. 2014).

A group of plant peptides that are processed from their pro-proteins by proteolytic cleavage and secreted upon herbivore attack, wounding or microbial infection were considered earlier to be secondary endogenous DAMP elicitors (Gust et al. 2017; Yamaguchi and Huffaker, 2011). Recently, these peptides were referred to as phytocytokines, since their release is not necessarily linked to tissue damage (Gust et al. 2017). They include Systemin identified in tomato (Pearce et al. 1991), which is recognized by its leucine rich repeat receptor-like kinase (LRR-RLK) SYR1 (Wang et al. 2018) and hydroxyproline-rich systemins (HypSys) of several Solanaceous plants (Pearce 2011). Other examples include the Plant Elicitor Peptides (PEPs) originally identified in *Arabidopsis* and recognized by two closely related LRR-RLKs, *At*PEPR1 and *At*PEPR2 (Bartels and Boller 2015; Krol et al. 2010). The *Arabidopsis* PAMP-Induced Peptides (PIPs), which are recognized by the LRR-RLK RLK7 to trigger immune responses in a manner similar to *At*PEP1 and bacterial flagellin-derived peptide Flg22, are also considered among the group of these phytocytokines (Hou et al. 2014). It was found that both PEP and PIP cooperatively amplify the immune responses triggered by Flg22 in *Arabidopsis* (Hou et al. 2014), and PEP amplifies HAMP-induced defense responses in rice (Shinya et al. 2018). Figure 1.1 presents a diagram summarizing plant innate immunity induced against insect herbivore.

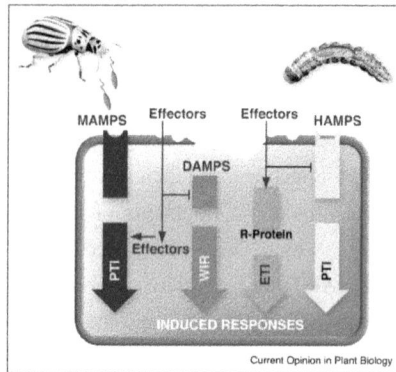

Figure 1.1: Model for Perception of Herbivore Elicitors.
HAMPs released from herbivorous insects as well as the MAMPs released from their microbial endosymbionts are recognized by their respective PRRs and trigger PTI. DAMPs also are recognized by cognate PRRs and induce wound-induced resistance (WIR). Herbivore and microbial effectors may suppress WIR and PTI. The microbial symbionts may be included in the oral secretions of some herbivore insects such as beetles (Chung et al. 2013) or may modify the expression of HAMPs or herbivore effectors, which can be recognized by plant R-proteins and trigger ETI (modified from Acevedo et al. 2015).

Increasing evidence suggests that the perception of elicitors from different sources such as HAMPs, MAMPs and DAMPs by their PRRs activate the same PTI responses via pathways that share the same intracellular signaling events but with differences in lag phases, amplitudes and kinetics (Choi and Klessig 2016; Ranf et al. 2011; Schmelz 2015). These intracellular signaling events include membrane depolarization, ion fluxes, intracellular Ca^{2+} influx, production of Reactive Oxygen Species (ROS), activation of the defense-associated Mitogen-Activated Protein Kinases (MAPKs), phytohormone biosynthesis and perception, activation of transcription factors, biosynthesis of defensive proteins and accumulation of defense-related metabolites (Choi and Klessig 2016; Schmelz 2015).

Recognition of effectors by plant R proteins induces ETI including intracellular signaling events that partially overlap with PTI with temporal and quantitative differences (Cui et al. 2015). ETI eventually induces the production of various antimicrobial proteins called Pathogenesis-Related (PR) proteins in and around the infected cell accompanied by the development of local Hypersensitive Response (HR; Fu and Dong 2013).

The key defense-related phytohormones involved in innate immune responses against insect herbivores include jasmonic acid (JA), ethylene (ET), and/or salicylic acid (SA; Schmelz, 2015). JA and sometimes ET orchestrate the downstream production of defense compounds locally and systemically that are effective against necrotrophic pathogens and chewing insects (Erb et al. 2012). On the other hand, SA plays an essential role in both local and systemic defenses against biotrophic pathogens and sap-sucking herbivores (Bastías et al. 2018; Vidhyasekaran 2015). The signaling pathways of these hormones intersect and can interact in an antagonistic or synergistic way allowing plants to fine tune their responses to specific biotic threats (Acevedo et al. 2015).

1.3. Herbivore- and Wound-Induced Defense Responses in Tomato

Like other higher plants, the cultivated tomato (*Solanum lycopersicum*) has trichomes on its stems and leaves representing one of the first defense lines against herbivorous insects (Yazaki et al. 2017). Tomato glandular trichomes contain a mixture of volatile terpenoids consisting mostly of monoterpenes and sesquiterpenes such as γ-terpinene, limonene, ß-caryophyllene, α-humulene and γ-elemene that act as repellent to these insects (Schilmiller et al. 2009).

Tomato also responds to herbivorous insects and mechanical wounding by induction of defense responses (Sun et al. 2011). One of the end products of these responses is the production of proteinase inhibitor I and II in wounded and distal unwounded leaves, which renders tomato leaf material indigestible for insects (Bosch et al. 2014a; Schaller and Ryan 1996). These defense responses are induced in tomato, like in other plants, by recognition of the released endogenous DAMPs and HAMPs of invading insects (Bowles 1998; Lortzing and Steppuhn 2016). On the other hand, wounding and herbivore attacks in tomato and some members of *Solanaceae* induce the release of a special phytocytokine called 'Systemin' (Pearce et al. 1991).

1.3.1. Discovery of Systemin

Tomato plants were used for many years as a model to study wound responses and the signals responsible for systemic defense response using PI-I and II as response markers (Bowles 1998). This was first reported by Green and Ryan in 1972, who demonstrated the systemic nature of the wound response. They showed that mechanical wounding or herbivore damage of tomato and potato leaves induce a rapid accumulation of PIs throughout the plants' tissues (Green and Ryan 1972). The pectic OGs enzymatically released from cell walls at the damage site were the first candidates to be proposed as systemic wound signals (Bishop et al. 1981). Few years later, Baydoun and Fry (1985) disproved that and suggested that the cell wall-derived oligosaccharides act locally and induce a secondary messenger to be the systemic signal. The search for the systemic wound signal lead in 1991 to the isolation and identification of a 18-mer peptide from tomato leaves that was able to induce the expression of PIs systemically when supplied to young leaves at a very low concentration (Pearce et al. 1991). A synthetic peptide with the same amino acid sequence as the isolated peptide triggered the same response (Pearce et al. 1991). Because this newly discovered 18-mer peptide was thought to be the systemic signal it was called 'Systemin', making it the first identified peptide hormone in plants (Pearce et al. 1991).

Like animal peptide hormones, Systemin is derived from a larger precursor protein called Prosystemin (Figure 1.2; Schaller and Ryan, 1996). This 200 amino acids-long protein is encoded by a single copy gene that consists of 11 exons and 10 introns, ten exons are organized as five homologous pairs with an unrelated sequence in the eleventh, encoding Systemin (McGurl et al. 1992; McGurl and Ryan, 1992). The accumulation of Prosystemin mRNA is wound-inducible in tomato leaves (McGurl et al. 1992). The shoot to root bidirectional translocation of this mRNA could be shown by heterologous shoot- and root-specific expression of Prosystemin in *Arabidopsis*, which was confirmed by grafting experiments (Zhang et al. 2018a). Prosystemin is compartmentalized in the cytosol and the nucleus of parenchyma cells of the vascular phloem in the leaves, various floral organs and in the bicollateral phloem bundles of petioles and stems of tomato (Narváez-Vásquez and Ryan 2004). Silencing of Prosystemin in transgenic tomato plants resulted in reduced defense protein accumulation and thus loss of resistance to insect herbivores (Orozco-Cardenas et al. 2006), whereas transgenic plants overexpressing Prosystemin exhibited high constitutive levels of defense proteins and increased resistance to insect attacks (Chen et al. 2005).

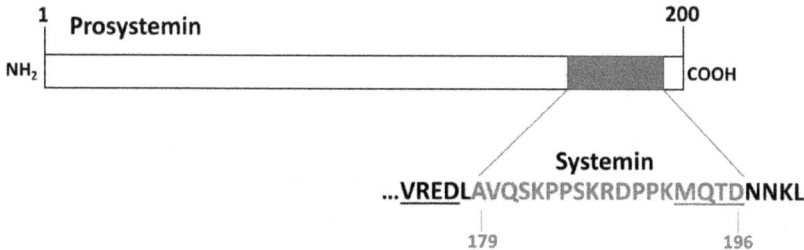

Figure 1.2: Primary Structure of Systemin and Prosystemin.
Schematic representation of the 200-amino acid Prosystemin precursor protein. The 18-amino-acid Systemin sequence is located near the C-terminus and shown in blue. Recognition motifs for maturation enzymes are underlined (adapted from Schaller and Ryan 1996; Beloshistov et al. 2018; Dreizler 2018).

Homologs of tomato Systemin have been identified in other members of the *Solanaceae* such as potato (*Solanum tuberosum*), black nightshade (*Solanum nigrum*), and bell pepper (*Capsicum annuum*; Constabel et al. 1998). It was found that Systemin homologs of potato and pepper were as active as that of tomato, whereas nightshade Systemin was 10-fold less active (Constabel et al. 1998). Tobacco however, also a member of the *Solanaceae*, lacks Prosystemin and does not respond to Systemin treatment (Ryan 2000).

In tomato, insect attack, pathogen infection or mechanical wounding trigger the cleavage of Systemin from its precursor protein Prosystemin (Zhai et al. 2017). Recently, Beloshistov et al. (2018) reported that the cleavage of Prosystemin at two Systemin-flanking aspartic acid residues can be accomplished *in vitro* by two phytaspases, *Sl*Phytaspase-1 (*Sl*Phyt-1) and *Sl*Phyt-2. This cleavage results in the production of Systemin with an extra leucine (Leu) residue at the amino terminus (Leu-Systemin). A Leu aminopeptidase A (LAP-A), which is involved in tomato wound responses (Fowler et al. 2009; Gu and Walling 2000), was suggested to perform the N-terminal truncation of Leu-Systemin to accomplish Systemin maturation during wounding (Beloshistov et al. 2018).

In an analysis of structure-function relationship, amino acid substitutions in the amino (N)-terminal half of the Systemin peptide were found to have little or no effect on its PI-inducing activity (Pearce et al. 1993). In contrast, substitution of Pro13 to Ala (A13) causes a major reduction in activity, while removal of the last amino acid Asp18, or substitution of Thr17 to Ala (A17), abolishes the activity (Pearce et al. 1993). Furthermore, it was found that the C-terminal tetrapeptide itself is able to induce *PI* gene expression indicating that the C-terminal part of Systemin is critical for its activity, although much higher concentration of this tetrapeptide than the native peptide was needed (Meindl et al. 1998; Pearce et al. 1993). Meindl et al. (1998) demonstrated that the first 14 N-terminal amino acids of Systemin antagonize peptide activity competitively, indicating that the N-terminal part of Systemin is important for its binding to the receptor. Similarly, the inactive A17 peptide has an antagonistic effect on Systemin perception (Felix and Boller, 1995; Meindl et al. 1998; Pearce et al. 1993). Pretreatment of plants with A17 prior to Systemin blocks the action of Systemin and prevents *PIs* gene expression (Pearce et al. 1993). Likewise, A17 has an antagonistic effect on Systemin perception when applied together with Systemin to *Solanum peruvianum* cell suspension culture (Felix and Boller 1995). This cell suspension culture is very sensitive to Systemin stimulation, which causes rapid and transient alkalization of the growth medium with a half-maximal activity at concentrations of ~100pM (Felix and Boller 1995). Recently, Chowdhury and Lahiri (2017) used extensive Molecular Dynamics (MD) simulation to demonstrate that Systemin has a tendency to adopt polyproline II (PPII) conformation in solution. Depending on *in silico* calculations they concluded that this conformation may be important for receptor binding (Chowdhury and Lahiri 2017).

Initially Systemin was thought to be the systemic signal that is transported across the vascular tissue to induce the wound response in distal tissues (Pearce et al. 1991; Ryan 2000; Schaller and Ryan 1996). However, a few years later, it could be shown that JA or a JA derivative such as methyl-jasmonate (MeJA) or JA-Ile are responsible for this and that Systemin functions locally to increase JA synthesis to the level required for the systemic response (Li et al. 2002, 2003).

1.4. Systemin Signaling Pathway

Systemin induces the expression of defense genes in tomato via the octadecanoid pathway to result in the production of JA as a powerful inducer of defense gene transcription (Sun et al. 2011). This involves a signal cascade that starts with binding of Systemin to its receptor followed by a series of signal transduction events (Ryan 2000; Schaller and Ryan 1996; Sun et al. 2011).

In 1999, a 160-kDa Systemin-binding protein was isolated from membranes of *Solanum peruvianum* suspension cultured cells and was claimed to be the Systemin receptor, SR160 (Scheer and Ryan 1999). However, SR160 was subsequently shown to be the tomato homolog of *Arabidopsis* Brassinosteroid Insensitive1 (*S*lBRI1; Scheer and Ryan, 2002), and *S*lBRI1 was found to be dispensable for Systemin-induced wound responses, ion fluxes, and defense genes expression (Holton et al. 2007; Lanfermeijer et al. 2007). Recently, a pair of tomato RLKs, Systemin Receptor1 (SYR1) and SYR2, were reported to be involved in Systemin perception with SYR1 as the high-affinity *bona fide* Systemin receptor, although its expression was not critical for local and systemic wound responses (Wang et al. 2018). On the other hand, a PEPR1 homolog called PEPR1/2 Ortholog Receptor-like Kinase1 (PORK1) was reported recently to play a crucial role in the Systemin signaling pathway in tomato (Xu et al. 2018). Silencing of PORK1 rendered the plants susceptible to herbivore attack and compromised the expression of jasmonates biosynthesis and *PI-II* genes in response to Systemin and wounding (Xu et al. 2018). This indicates that possibly the interaction of more than one receptor might be involved in Systemin perception (Xu et al. 2018).

Perception of Systemin by its receptor causes rapid depolarization of the plasma membrane (PM) potential associated with rapid alkalization of the apoplast and growth medium of *S. peruvianum* cells as a result of PM H^+-ATPase inactivation (Felix and Boller 1995; Schaller and Oecking 1999). Membrane depolarization is accompanied by H^+ and Ca^{2+} influx and K^+ efflux, which lead to a rapid increase in cytoplasmic Ca^{2+} concentrations (Dombrowski and Bergey 2007; Schaller and Oecking 1999). Consistent with a role for Ca^{2+} in Systemin signaling a transient systemic increase in the mRNA and protein levels of calmodulins such as CaM6 was reported after wounding and Systemin treatment (Bergey and Ryan 1999; Stanković and Davies 1997; Xu et al. 2018). Furthermore, Dombrowski and Bergey (2007) reported that increased cytosolic Ca^{2+} concentration is critical for the activation of the octadecanoid pathway responsible for JA biosynthesis.

Perception of Systemin also induces the production of ROS, which act as secondary messengers for the activation of defense gene expression (Orozco-Cardenas and Ryan 1999; Orozco-Cardenas et al. 2001). In addition, ROS have direct antimicrobial activity, and they contribute indirectly to defense by strengthening of the cell wall through oxidative cross-linking of glycoproteins (Boller and Felix 2009; Muthamilarasan and Prasad 2013). Orozco-Cardenas et al. (2001) demonstrated that supplying tomato seedlings with inhibitors of NADPH oxidase, a membrane-bound enzyme responsible for ROS production, diminished the expression of defense genes, namely *PI-I* and *II* in response to wounding and Systemin application. In contrast, inhibition of NADPH oxidase did not inhibit the wound-inducible expression of *Prosystemin*, *LOX* and *Allene Oxide Synthase* (*AOS*) genes, which are associated with the octadecanoid signaling pathway (Orozco-Cardenas et al. 2001). This observation suggested that ROS-mediated induction of the later-expressed defensive genes is independent of the early-expressed signaling-related genes (Orozco-Cardenas et al. 2001). Silencing of two NADPH oxidases (Respiratory Burst Oxidase Homologs, RBOHs) in tomato compromised wound-induced systemic expression of *PI-II*, confirming that ROS intermediates supplied by RBOH are required for the wound response (Sagi et al. 2004).

Like many stress-related signaling pathways, Systemin perception induces the activation of MAPKs (Holley et al. 2003). In *S. peruvianum* cell cultures, Systemin stimulation induces the activation of two closely related MAPKs, namely MPK1 and MPK2 (Holley et al. 2003). Kandoth et al. (2007) showed that co-silencing of MPK1 and MPK2 in transgenic tomato plants overexpressing *Prosystemin* reduces MPK1/2 kinase activity, JA biosynthesis and expression of defense genes in response to wounding and herbivore attack. The full defense response was restored after application of MeJA indicating that MPK1 and MPK2 most likely function upstream of JA biosynthesis (Kandoth et al. 2007).

Biosynthesis of the plant stress hormone ethylene (ET) is reported to be induced after Systemin perception (Wang et al., 2002). In *S. peruvianum* cell cultures Systemin causes a rapid increase in 1-Aminocyclopropane1-Carboxylate Synthase (ACS) activity, the rate-limiting enzyme in ET biosynthesis (Felix and Boller 1995). Similarly, wounding and application of Systemin or JA triggers a rapid and transient ET burst in tomato seedlings (O'Donnell et al. 1996). Supporting a role for ET in the wound response, the accumulation of PIs after application of JA, Systemin or wounding was found to be impaired in tomato plants blocked in ET synthesis or signaling (O'Donnell et al. 1996). On the other hand, JA levels after wounding were significantly reduced in these plants as compared to wild-type plants but were higher than in unwounded plants (O'Donnell et al. 1996). This indicates that ET is important for full induction of JA production in response to Systemin and wounding, and a cross-talk between both hormones is important for the induction of *PI* expression (O'Donnell *et al.*, 1996; Sivasankar *et al.*, 2000). Interestingly, in other plants species it was reported that ROS may act as a signal for ET formation in wounded leaves (Steinite et al., 2004; Wang et al., 2002; Watanabe and Sakai, 1998).

Upon wounding, Phospholipase A (PLA) activity increases rapidly and systemically in tomato leaves (Narvaez-Vasquez et al. 1999). PLA was proposed to act on chloroplast membranes to release linolenic acid, which is subsequently converted via the octadecanoid pathway to JA (Dombrowski and Bergey 2007; Farmer and Ryan 1992). Narvaez-Vasquez et al. (1999) demonstrated that Systemin is required for part of the systemic PLA activation. Additionally, other enzymes of the octadecanoid pathway, including for example AOS, are induced in tomato leaves upon wounding or Systemin application (Sivasankar et al. 2000; Wasternack et al. 2006). Tomato mutants that are compromised in their ability to accumulate JA or in JA perception are unable to accumulate defense proteins in response to mechanical wounding, herbivore feeding or Systemin application indicating that octadecanoid metabolism plays an essential role in the signal transduction responsible for the activation of anti-herbivore plant defenses (Howe et al. 1996; Li et al. 2001; Yan et al. 2013)

JA biosynthesis starts with the conversion of linolenic acid released by PLA into (13-S)-hydroperoxy linolenic acid (13-HPOT) by the activity of LOX (Heitz et al. 1997). Then AOS converts 13-HPOT into an unstable allene oxide that is further processed by AOC (Froehlich et al. 2001). In the AOC-catalyzed reaction cis-(+)-OPDA is formed which carries the enantiomeric structure at the cyclopentenone ring of the final product (+)-7-iso-JA (Ziegler et al. 2000). The previous reactions take place in the stroma of plastids, but the following reactions occur in peroxisomes (Froehlich et al. 2001; Strassner et al. 2002a; Wasternack et al. 2006). Thus, OPDA is exported from the plastid and imported into the peroxisomes either through the peroxisomal ABC transporter comatose (CTS), or by an alternative mechanism like ion trapping (Theodoulou et al. 2006). Here OPDA is reduced by the cis-(+)-OPDA-specific OPDA-Reductase 3 (OPR3) to 3-oxo-2-(cis-2'-pentenyl)-cyclopentane-1-octanoic acid (OPC-8; Strassner et al. 2002b) followed by three cycles of ß-oxidation involving Acyl-CoA Oxidase (ACX), Multifunctional Proteins (MFP) and 3-Ketoacyl-CoA Thiolase (KAT; Li et al. 2005; Miersch and Wasternack 2000). The released JA may be metabolized by methylation at the carboxylic acid group to form MeJA, the volatile defense signal, or by conjugation to amino acids such as isoleucine to form the active hormone JA-Ile (Staswick and Tiryaki 2004).

The orchestration of events induced by Systemin triggers a wide transcriptome reprogramming, inducing the expression of genes involved in defense responses (Coppola et al. 2015). This transcriptome reprogramming is mediated by an array of transcription factors mainly activated by JA and ET (Du et al. 2017). Examples of such transcription factors are WRKY40, MYC2 and JA2L as well as proteins of bHLH, NAC, ERF and MYB families (Coppola et al. 2015; Du et al. 2017). These transcription factors induce the expression of not only the defense genes, but also the genes expressing the proteins involved in the signaling cascade such as Prosystemin, RLKs involved in the signaling cascade and the enzymes of the octadecanoid pathway, which work as positive feedback to amplify the signal (Boller and Felix 2009; Du et al. 2017; Ryan 2000). Figure 1.3 summarizes the Systemin signaling cascade as described in the text above.

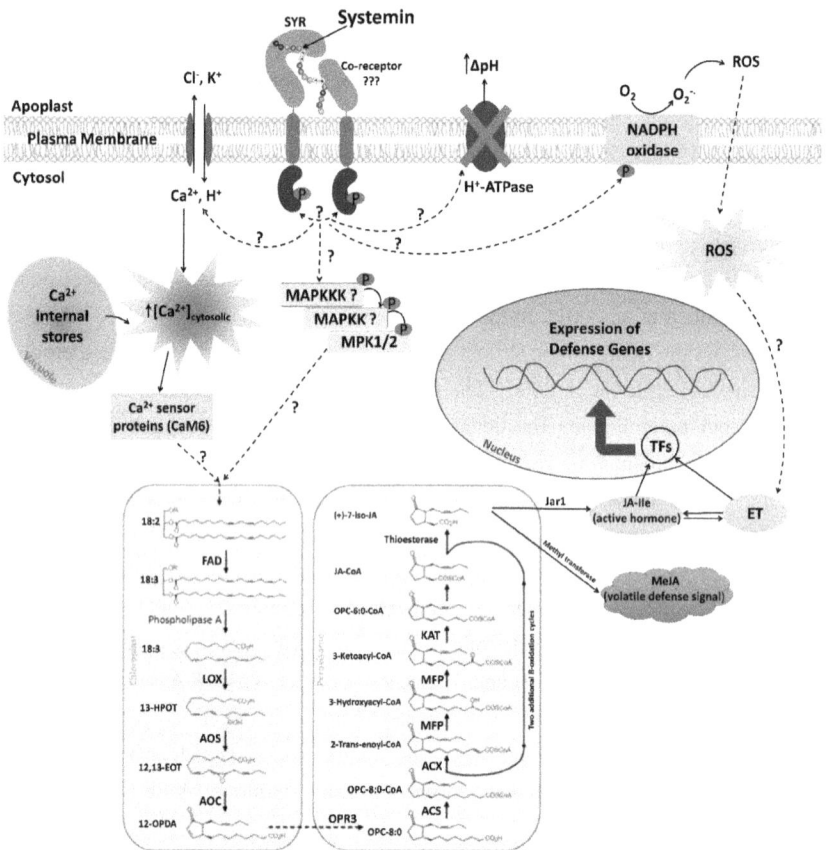

Figure 1.3: The Systemin Signaling Cascade.
Wounding or herbivore attack induces the processing of Systemin from its precursor Prosystemin (not shown). The Systemin oligopeptide binds to its receptor to activate defense signaling cascade possibly via the interaction with a co-receptor, which is still elusive. This interaction causes transphosphorylation of the kinase domains of the receptors, which then by an unknown way induces membrane

depolarization associated with H^+ and Ca^{2+} influxes and K^+ efflux and alkalization of the apoplast because of H^+-ATPase inactivation. This leads to an increase of cytosolic Ca^{2+} concentration, accumulation of ROS, and activation of a MAPK cascade. In an unknown way these cellular events lead to the biosynthesis and the release of JA through the octadecanoid pathway. Jasmonates transport the defense signal to the other parts of the plant. In crosstalk with ET transcription factors (TFs), the activated TFs are responsible for the expression of defense genes. Adapted from (Hohmann et al. 2017; Kandoth et al. 2007; Ryan 2000; Schaller 1999; Schaller and Stintzi 2008; Wang et al. 2002; Xu et al. 2018; Zhai et al. 2017).

1.5. Applications of Phosphoproteomics in Studying Plant Signaling Pathways

One of the abundant biochemical modifications involved in cellular signaling cascades in most eukaryotic cells is protein phosphorylation (Mumby and Brekken 2005). This reversible posttranslational modification mastered by protein kinases and phosphatases is a primary mean of modulating the activity, subcellular localization and half-life of substrate proteins in response to a stimulus (Invergo and Beltrao 2018). Cascades of phosphorylation events on protein kinases and phosphatases form the basis of many intracellular signaling pathways, in which the activity of these proteins themselves is often regulated by phosphorylation (Cheng et al. 2011; Luan 2003). This is the case also for most plant defense response signaling pathways, which are repressed if the plant is treated with a general inhibitor of serine/ threonine kinases such as K-252a, indicating the essential role of these enzymes in regulating the signal transduction (Peck 2003; Pedley and Martin 2005). Therefore, the analysis of signaling pathways in plants has often focused on protein kinases and phosphatases (De La Fuente Van Bentem et al. 2006; Zhang et al. 2015).

For quantifying the phosphorylation state of proteins, traditional biochemical approaches as well as genetic analyses of phosphoproteins, and of the kinases and phosphatases that modify them, have provided a wealth of information about signaling pathways (Invergo and Beltrao 2018). The limitations of such approaches are that only a limited number of proteins can be included in any given study, and that the complexity of protein phosphorylation in the context of signaling networks is not taken into account (Invergo and Beltrao 2018; Mumby and Brekken 2005). The availability of well-annotated genome databases and advancements in mass spectrometry (MS) have paved the way to study many phosphoproteins and phosphorylation sites at once using a phosphoproteomic approach (Invergo and Beltrao 2018; Mumby and Brekken 2005; Zhang et al. 2015).

Phosphoproteomics enables the identification and quantification of phosphorylated proteins at proteome scale, and the acquisition of a thorough view of the degree and dynamics of protein phosphorylation in entire signaling networks (Mumby and Brekken 2005; Xing and Laroche 2011). This technique relies on MS-based identification and quantification of phosphopeptides enriched from complex biological samples (Nakagami et al. 2012). The most extensively used methods for phosphopeptide enrichment depend on the affinity interaction between metal ions such as Fe^{3+}, Ga^{3+} and Ti^{4+} and phosphate groups of the phosphopeptides (Li et al. 2016). Such methods include Immobilized Metal Affinity Chromatography (IMAC) and Metal Oxide Affinity Chromatography (MOAC) that are most commonly used due to their selectivity and sensitivity (Li et al. 2016; Salovska et al. 2012)

Large scale phosphoproteomics has proved to be a powerful technique to identify key players in plant signaling networks. For example, time course large scale phosphoproteomics enabled the identification of new phosphorylation sites in the PM H^+-ATPase in response to sucrose (Niittylä et al. 2007). The data gained from this experiment paved the way for the identification of SIRK1, a novel receptor kinase involved in the regulation of aquaporins in response to sucrose (Wu et al. 2013). Using a similar approach, a comprehensive picture of the early and late responses induced by nitrogen resupply to starved *Arabidopsis* seedlings was obtained (Engelsberger and Schulze 2012). Additionally, the receptor of the Rapid Alkalinization Factor (RALF) peptide, Feronia, as well as its downstream target, the PM H^+-ATPase, were successfully identified using phosphoproteomics (Haruta et al. 2014). Recently, a systemic approach combining metabolic labeling and phosphoproteomics was used to identify early signaling events in response to salt and oxidative stresses (Chen and Hoehenwarter 2015). This study provided an evidence for a connection between early signaling events in salt and oxidative stress responses, which regulates the state transition of photosynthesis and the rearrangement of primary metabolism (Chen and Hoehenwarter 2015).

1.6.   Aims of this work

Plant innate immune responses triggered by PRR-induced signaling cascades involve the regulation of protein activity by phosphorylation (Boller and Felix 2009). For example, the transient inactivation of the PM H^+-ATPase and activation of MAPKs are common features of perception of various immunity elicitors such as Systemin, Flg22 and OGs (Elmore and Coaker 2011; Holley et al. 2003; Nühse et al. 2000; Schaller and Oecking 1999; Thain et al. 1995). Some of these transient changes in the activity of signaling proteins contribute to the signal transduction and expression of defense responses (Boller and Felix 2009). A clear example is the activation of NADPH oxidases, which produce ROS that stimulate the expression of *ACS* inducing the production of ET after wounding in winter squash (Watanabe and Sakai 1998). In contrast, changes in activity of other key players may negatively affect the signaling cascade and attenuate its response, for example MKKK7, which negatively regulates flagellin-triggered signaling in *Arabidopsis* by direct modulation of the FLS2 receptor complex (Mithoe et al. 2016).

The Systemin receptor SYR1 is a typical RLK with a ligand-binding LRR ectodomain and an intracellular kinase domain (Wang et al. 2018). Like for other PRR signaling pathways, intracellular signaling is initiated by activation of the kinase upon binding of Systemin to SYR1 (Macho and Zipfel 2014; Wang et al. 2018). Since the identification of Systemin almost 30 years ago, the specific phosphorylation events that initiate signaling, the interaction with downstream substrates, and subsequent signaling transduction events that lead to JA biosynthesis and eventually defense genes expression remain unknown. Advanced mass spectrometric techniques that detect and quantify changes in the phosphoproteome provide a chance to identify vital transient phosphorylation events involved in this signaling cascade.

The first aim of this work was to identify the early intracellular (de)phosphorylation events induced by Systemin, the involved kinases and phosphatases, as well as their potential substrates. To capture these transient events with high confidence a time-course large-scale phosphoproteomics approach was followed. *In vitro* assays were used to detect the transient inactivation or activation of some proteins involved in the Systemin signaling cascade.

Perception of some plant immunity elicitors such as chitin was found to be mediated by a complex of multiple LysM receptors with different chitin affinities (Cao et al. 2014;

Petutschnig et al. 2010). Similarly, as reported by Wang et al. (2018) and Xu et al. (2018), several RLKs seem to be involved in Systemin perception and signal transduction. The second aim of this work was to study possible positive or negative contributions of Systemin-induced RLKs identified by the phosphoproteomics approach, as well as of the known SYR1, SYR2 and PORK1 receptors, to the Systemin signaling pathway using a loss-of-function approach.

2. Materials and Methods

2.1. Materials

2.1.1. Chemicals and Consumables

Plastic consumable materials like pipette tips, reaction tubes and petri-dishes were purchased from Sarstedt (Nümbrecht, Germany). All chemicals and antibiotics were purchased from Roth (Karlsruhe, Germany), Merck (Darmstadt, Germany), Fluka (Buchs, Switzerland), Sigma-Aldrich (Taufkirchen, Germany), Serva (Heidelberg, Germany) and Duchefa (Haarlem, The Netherlands).

2.1.2. Enzymes

Restriction enzymes and appropriate buffers, T4-DNA ligase and alkaline phosphatase (FastAP) were obtained from Thermo Fisher Scientific (Waltham, United States) and New England Biolabs (Ipswich, United States). All enzymes were used according to the manufacturer's instructions with the appropriate buffers

2.1.3. Antibiotics

Table 2.1: Used antibiotics. The stock solutions were stored at -20°C.

Antibiotic	Stock concentration and solvent
Carbenicillin	100 mg/ml in ddH$_2$O
Chloramphenicol	34 mg/ml in 70% (v/v) EtOH
Gentamycin	50 mg/ml in ddH$_2$O
Kanamycin	50 mg/ml in ddH$_2$O
Spectinomycin	100 mg/ml in ddH$_2$O
Tetracyclin	12,5 mg/ml in 70% (v/v) EtOH

| Ticarcillin | 250 mg/l used as powder weighed and dissolved directly in medium after autoclave. The powder was stored at 4°C. |
| Vancomycin | 500 mg/l used as powder weighed and dissolved directly in medium after autoclave. The powder was stored at 4°C. |

2.1.4. Chemicals used for Plant Tissue Culture

Table 2.2: Chemicals used to prepare tissue culture medium. The stock solutions were stored at -20°C.

Hormone	Stock concentration and solvent
N-Benzyladenine	40 mg/ml in 1 N NaOH
Benzylaminopurine (BAP)	10 mg/ml in 1 N NaOH
Indole acetic acid (IAA)	1 mg/ml in 96% (v/v) EtOH
Naphthalene acetic acid (NAA)	10 mg/ml in 96% (v/v) EtOH
Trans-Zeatin	10 mg/ml in 1 N NaOH
Acetosyringone	400 mM in DMSO
Thiamin	9.6 mg/ml in ddH$_2$O
Pyridoxine	1 mg/ml in ddH$_2$O
Nicotinic acid	1 mg/ml in ddH$_2$O

2.1.5. Other Chemicals

Table 2.3: Other Chemicals Used. The stock solutions were stored at -20°C.

Chemical	Stock concentration and solvent
IPTG	100 mM in ddH2O
X-Gal	20 mg/ml in DMSO

RNAse	10 mg/ml in TE buffer heated for 10 min at 85°C
Ouabain	0.1 mM in ddH_2O
NaN₃	100 mM in ddH_2O
Concanamycin A	12 µM in DMSO
Orthovanadate	10 mM in H^+-ATPase assay reaction buffer (section 2.4.10), pH was adjusted to 10 with NaOH
Lys-C	0.5 µg/µl in ddH_2O (HPLC grade)
Trypsin	0.5 µg/µl in 50 mM acetic acid (HPLC grade)

2.1.6. Media

2.1.6.1. Bacterial media

LB Medium:

LB Broth 20 g/l, for solid medium 0.75% (w/v) agar.

SOC Medium:

20 g/l Bacto-Tryptone, 5 g/l yeast-extract, 10 mM NaCl, 2.5 mM KCl ,10 mM $MgCl_2$, 10 mM $MgSO_4$, 20 mM Glucose, pH 6.8.

2.1.6.2. *S. peruvianum* Media

Nover-Medium:

3% (w/v) sucrose, 4.41 g/l MS-Salts (Murashige and Skoog + Nitsch Vitamins, Duchefa), 0.34 g/l KH_2PO_4, 5 mg/l naphthalene acetic acid (NAA), 2 mg/l N-Benzyl adenine, pH 5.5 adjusted with 2N KOH. Kanamycin with final concentration of 75 mg/l was added for transgenic cell suspension cultures. For solid medium (used in section 2.9) 0.6% agar (w/v) was used.

2.1.6.3. Media for Tomato Transformation

Germination Medium:

3% (w/v) sucrose, 4.4 g/l MS-Salts (Murashige and Skoog basal salts, Duchefa), pH 5.8 adjusted with 0.5 N KOH, 0.6% (w/v) agar.

Conditioning Medium:

3% (w/v) sucrose, 4.4 g/l MS-Salts (Murashige and Skoog basal salts, Duchefa), 1 mg/l NAA, 0.1 mg/l Benzylaminopurine (BAP) pH 5.8 adjusted with 0.5 N KOH, 0.6% (w/v) agar.

Co-cultivation Medium:

3% (w/v) sucrose, 4.4 g/l MS-Salts (Murashige and Skoog basal salts, Duchefa), pH 5.8 adjusted with 0.5 N KOH. 200 µM Acetosyringone was supplemented to the medium before use.

Selection Medium:

3% (w/v) sucrose, 4.4 g/l MS-Salts (Murashige and Skoog basal salts, Duchefa), 9.6 mg/l thiamin, 1 mg/l nicotinic acid, 1 mg/l pyridoxine, 1 mg/l trans-zeatin, 35 (50, 100) mg/l kanamycin, 250 mg/l ticarcillin, pH 5.8 adjusted with 0.5 N KOH, 0.6% (w/v) agar.

Rooting Medium:

3% (w/v) sucrose, 4.4 g/l MS-Salts (Murashige and Skoog basal salts, Duchefa), 9.6 mg/l thiamin, 1 mg/l nicotinic acid, 1 mg/l pyridoxine, 0.1 mg/l indole acetic acid, 20 mg/l kanamycin, 500 mg/l vancomycin, pH 5.8 adjusted with 0.5 N KOH, 0.6% (w/v) agar.

2.1.7. Peptides

Table 2.4: used peptides. All peptides were dissolved in ddH$_2$O to make a stock solution of 1 mg/ml and stored at -20°C.

Systemin (AVQSKPPSKRDPPKMQTD) Purity: >95 % (Pepmic Co., Ltd, Suzhou, China)

A17 (AVQSKPPSKRDPPKMQAD) Purity: >95 %
(Pepmic Co., Ltd, Suzhou, China)

LHA1T955 (GLDIETIQQSYTV) Purity: >95 %
(Pepmic Co., Ltd, Suzhou, China)

LHA1T955(ph) (GLDIETIQQSYT(ph)V) Purity: >95 %
(Pepmic Co., Ltd, Suzhou, China)

Flg22 (QRLSTGSRINSAKDDAAGLQIA) Purity: >95 %
(GenScript Biotech Corp., Piscataway, United States)

2.1.8. Primers

All primers were designed using CLC Main Workbench software v. 8.1 (Qiagen, Aarhus, Demark) and analyzed by Beacon Designer™ Free Edition (Premier Biosoft). All primers were obtained from Eurofins Genomics (Ebersberg, Germany), resuspended in Tris/EDTA (TE) buffer (10 mM Tris/HCl pH 8.0, 1 mM EDTA) at a concentration of 100 μM and stored at -20°C.

Table 2.5: Primers used for cloning and expression of MPK2, PLL5 phosphatase domain (PD), and GHR1 intracellular domain (ID).

Restriction sites are underlined, start and stop codons are in bold and the His$_6$-tag is highlighted.

* These primers were used to introduce two mutations (D216G/E220A) rendering MPK2 constitutively active. The mutated codons are marked in red.

Primer	Sequence (5' - 3')	Template
14420*Nco*I1F	ATTA<u>CCATGG</u>CAGATGGTTCAGCTCCG	Solyc08g014420 (MPK2)
14420D216G/E220AR1*	GTCACAACATATGCGGTCATAAAGCCAGTTTCAGAAG	Solyc08g014420 (MPK2)

14420D216G/E220AF2*	CTTCTGAAACTGGCTTTATGACCGCA TATGTTGTGAC	Solyc08g014420 (MPK2)
14420*Xho*I394R	ATAT<u>CTCGAG</u>CATGTGCTGGTATTCG GGATT	Solyc08g014420 (MPK2)
76100*Nco*I241F	ATA<u>CCATGG</u>CACATCACCATCACCAT CACTTCAGCAGTGAGTGTAGTTTG	Solyc06g076100 (PLL5 homolog)
76100*Xho*I708R	AATT<u>CTCGAG**TTA**</u>TGCACTGGATCTC CATATTC	Solyc06g076100 (PLL5 homolog)
70000*Nde*I624F	ATTA<u>CATATG</u>CATCACCATCACCATC ACCGAGCTTCAAGGAAGCGTC	Solyc02g070000 (GHR1)
70000*Xho*I1024R	TTAA<u>CTCGAGCTA</u>TATAGACGAAAG GTCCTC	Solyc02g070000 (GHR1)

Table 2.6: Primers used for qRT-PCR.

Primer	Sequence (5' - 3')	Template
9650F(2547)	CAACAACGGTCATTCTTCATCAACT	Solyc01g109650
9650R(2628)	TCTCTCTACATCCTGCTCCTCCAA	Solyc01g109650
5010F(2229)	ACCCTTCGCTGGTTCTGATTG	Solyc07g005010
5010R(2320)	CCTTCATTCTCTCTGTCCTCCATTC	Solyc07g005010
66490F(2473)	TGAGTGAAGTGATTGGAGTGATGAG	Solyc08g066490
66940R(2577)	AAATGGTTGGTGAACAAGAAGAATG	Solyc08g066490
91400F(596)	CAAATGAGTTAGGAGATTGTGGGAAT	Solyc09g091400
91400R(678)	CAAAGAAACAGGCAAAGAACCAGA	Solyc09g091400
23950F(1923)	TGTTGTTTGTGTGGTTGTTCTTGTTG	Solyc02g023950
23950R(2038)	TTTGACGGCGGATGTCTTTGT	Solyc02g023950
70000F(1564)	CCCTCTGCTACACCCAACCTC	Solyc02g070000 (GHR1)
70000R(1646)	AACAAACCAAACCCATCAGGAAA	Solyc02g070000

24

		(GHR1)
81940F(839)	CCTACCTCAGCAACCAAATCCAA	Solyc08g081940
81940R(967)	CAACAAACAACCATCAAGAAACCAA	Solyc08g081940
83210F(969)	TACACCCTCAGCACCAGTAAATCAA	Solyc09g083210 (LYK4 homolog)
83210R(1122)	AACATCACCATACCCTTCTTTCTTCC	Solyc09g083210 (LYK4 homolog)
91840F(193)	AACCAAACAGTAGGTGGGCGTTAGA	Solyc02g091840
91840R(357)	CTGCGGGTCATCGGTAATGG	Solyc02g091840
63000F(3291)	ATCGCAAGTGTAAAGGAGAGGTGAG	Solyc07g063000 (PSKR2)
63000R(3460)	GAAACAAAGTGAACCAAACAAAGCA	Solyc07g063000 (PSKR2)
36330F(878)	GCATTCTTTCCTATCCCTCCTCCAT	Solyc12g036330 (LRK10L1.2)
36330R(1080)	TCATACAAACGCTTAACAGCAACGA	Solyc12g036330 (LRK10L1.2)
68300F(1273)	GGTCACAGTCTAACATATTGGGTGC	Solyc02g068300
68300F(1491)	ATCTTGCCCTTTGATGCTTTCCCTA	Solyc02g068300
PI2-161	GGATATGCCCACGTTCAGAAGGAA	Solyc03g020080 (PI-II)
PI2-418R	AATAGCAACCCTTGTACCCTGTGC	Solyc03g020080 (PI-II)
ToEFa-1288F	AGCCCATGGTTGTTGAGACCTTTG	Solyc06g005060 (Elongation Factor 1α)
ToEFa-1478R	TTCGAAACACCAGCATCACACTGC	Solyc06g005060 (Elongation Factor 1α)
Ubi3-230F	CTCTTGCCGACTACAACATCCA	Solyc01g056940 (Ubiquitin)

Ubi3-451R	AGCACCGCACTCAGCATTA	Solyc01g056940 (Ubiquitin)
ACT-816F	CACTACTGCTGAACGGGAAAT	Solyc03g078400 (Actin)
ACT-951R	CTGTCCATCTGGCAACTCATAG	Solyc03g078400 (Actin)

Table 2.7: Primers used for CRISPR/Cas9 constructs.

The single guide RNA (sgRNA) sequences are in bold. The *BsaI* recognition site for golden gate cloning is underlined. Recognition sites of restriction enzymes in sgRNA sequences are highlighted. The 5' G required for transcription initiation is small letter.

Primers with an ID that ending on F0 or R0 were diluted 20 times more than BsF or BsR primers (Xing et al. 2014).

Primer ID	Sequence (5' - 3')	Target Gene (Gene ID; Cas cleavage site)	Restriction enzyme for sgRNA	Template
DT824501-BsF	ATATATGGTCTCGATTg AGGGATGCCACCAGT GAAAGTT	Solyc03g082450 (SYR2; 305)	*BtsI:MutI*	pCBC-DT1T2
DT824501-F0	TgAGGGATGCCACCA GTGAAAGTTTTAGAGC TAGAAATAGC	Solyc03g082450 (SYR2; 305)	*BtsI:MutI*	pCBC-DT1T2
DT824502-R0	AACACGAGCTGACAG GACCTCTcAATCTCTT AGTCGACTCTAC	Solyc03g082450 (SYR2; 589)	-	pCBC-DT1T2
DT824502-BsR	ATTATTGGTCTCGAAA CACGAGCTGACAGGA CCTCTcAA	Solyc03g082450 (SYR2; 589)	-	pCBC-DT1T2
DT824701-BsF	ATATATGGTCTCGATTg CGGTAACTTGCCGGT CAATGTT	Solyc03g082470 (SYR1; 1253)	*HpaII*	pCBC-DT1T2

26

DT824701-F0	TgCGGTAACTTGCCG GTCAATGTTTTAGAGC TAGAAATAGC	Solyc03g082470 (SYR1; 1253)	*Hpall*	pCBC-DT1T2
DT824702-R0	AACCGTTTGGTCTCC CTTGATCcAATCTCTT AGTCGACTCTAC	Solyc03g082470 (SYR1; 288)	*Bsal*	pCBC-DT1T2
DT824702-BsR	ATTATTGGTCTCGAAA CCGTTTGGTCTCCCT TGATCcAA	Solyc03g082470 (SYR1; 288)	*Bsal*	pCBC-DT1T2
DT1096501-BsF	ATATATGGTCTCGATTg CTGGGATTCGTTGCA AGAAGTT	Solyc01g109650 (LRRXIV; 197)	*HpyCH4V*	pCBC-DT1T2
DT1096501-F0	TgCTGGGATTCGTTGC AAGAAGTTTTAGAGCT AGAAATAGC	Solyc01g109650 (LRRXIV; 197)	*HpyCH4V*	pCBC-DT1T2
DT1096502-R0	AAcCCAACTCGAGTA CGTTTAACAATCTCTT AGTCGACTCTAC	Solyc01g109650 (LRRXIV; 248)	*Xhol*	pCBC-DT1T2
DT1096502-BsR	ATTATTGGTCTCGAAAc CCAACTCGAGTACGT TTAACAA	Solyc01g109650 (LRRXIV; 248)	*Xhol*	pCBC-DT1T2
DT832101-BsF	ATATATGGTCTCGATTg GTTACAAGAACACGT GATGGTT	Solyc09g083210 (LYK4; 296)	*Pmll*	pCBC-DT1T2
DT832101-F0	TgGTTACAAGAACAC GTGATGGTTTTAGAGC TAGAAATAGC	Solyc09g083210 (LYK4; 296)	*Pmll*	pCBC-DT1T2
DT832102-R0	AAcCAAGATTCAGAG TAACCCCCAATCTCTT AGTCGACTCTAC	Solyc09g083210 (LYK4; 601)	-	pCBC-DT1T2
DT832102-BsR	ATTATTGGTCTCGAAAc CAAGATTCAGAGTAA CCCCCAA	Solyc09g083210 (LYK4; 601)	-	pCBC-DT1T2
DT630001-BsF	ATATATGGTCTCGATTg GAACCTAATTGCTGT AAATGTT	Solyc07g063000 (PSKR2; 196)	-	pCBC-DT1T2

DT630001-F0	TgGAACCTAATTGCTG TAAATGTTTTAGAGCT AGAAATAGC	Solyc07g063000 (PSKR2; 196)	-	pCBC-DT1T2
DT630002-R0	AAcATCTCTCCAAGGA CTGTGACAATCTCTTA GTCGACTCTAC	Solyc07g063000 (PSKR2; 301)	-	pCBC-DT1T2
DT630002-BsR	ATTATTGGTCTCGAAAc ATCTCTCCAAGGACT GTGACAA	Solyc07g063000 (PSKR2; 301)	-	pCBC-DT1T2
DT363301-BsF	ATATATGGTCTCGATTg CTGGTAAGCCACAGT AGGAGTT	Solyc12g036330 (LRK10L1.2; 165)	-	pCBC-DT1T2
DT363301-F0	TgCTGGTAAGCCACA GTAGGAGTTTTAGAGC TAGAAATAGC	Solyc12g036330 (LRK10L1.2; 165)	-	pCBC-DT1T2
DT363302-R0	AAcCGAAAGAATGGT GAGTGGCCAATCTCTT AGTCGACTCTAC	Solyc12g036330 (LRK10L1.2; 469)	-	pCBC-DT1T2
DT363302-BsR	ATTATTGGTCTCGAAAc CGAAAGAATGGTGAG TGGCCAA	Solyc12g036330 (LRK10L1.2; 469)	-	pCBC-DT1T2
DT700001-BsF	ATATATGGTCTCGATTg CCAATCTCTGGTGGT AAAGGTT	Solyc02g070000 (GHR1; 352)	-	pCBC-DT1T2
DT700001-F0	TgCCAATCTCTGGTGG TAAAGGTTTTAGAGCT AGAAATAGC	Solyc02g070000 (GHR1; 352)	-	pCBC-DT1T2
DT700002-R0	AAcAATCCCTGATACT ATTTCACAATCTCTTA GTCGACTCTAC	Solyc02g070000 (GHR1; 425)	-	pCBC-DT1T2
DT700002-BsR	ATTATTGGTCTCGAAAc AATCCCTGATACTAT TTCACAA	Solyc02g070000 (GHR1; 425)	-	pCBC-DT1T2
1598-Sl03g123860-BsF	ATATATGGTCTCGATTg TATCAGTGAATAAGC TTTCGTT	Solyc03g123860 (PORK1; 1598)	*HindIII*	pCBC-DT1T2

1598- Sl03g123860-F0	TgTATCAGTGAATAAG CTTTCGTTTTAGAGCT AGAAATAGC	Solyc03g123860 (PORK1; 1598)	*HindIII*	pCBC-DT1T2
DT0-BsR2	ATATTATTGGTCTCAA TCTCTTAGTCGACTCTA CCAAT	-	-	pCBC-DT1T2
1646- Sl03g123860-BsF	ATATTATTGGTCTCAA GATTgGTACGACAGAT TTAAACCCGTT	Solyc03g123860 (PORK1; 1646)	*DraI*	pCBC-DT2T3
1646- Sl03g123860-F0	TgGTACGACAGATTTA AACCCGTTTTAGAGCT AGAAATAGC	Solyc03g123860 (PORK1; 1646)	*DraI*	pCBC-DT2T3
1771- Sl03g123860-R0	AAcAGCTTCCAAACG CGGATGGCAATCACT ACTTCGTCTCTAACCA T	Solyc03g123860 (PORK1; 1771)	*HindIII*	pCBC-DT2T3
1771- Sl03g123860- BsR	ATTATTGGTCTCGAAAc AGCTTCCAAACGCGG ATGGC	Solyc03g123860 (PORK1; 1771)	*HindIII*	pCBC-DT2T3

Table 2.8: Primers used for colony PCR of CRISPR/Cas9 constructs (Xing et al. 2014).

Constructs	Sequence (5' - 3')
For two sgRNA constructs	U6-26p-F: TGTCCCAGGATTAGAATGATTAGGC U6-29p-R: AGCCCTCTTCTTTCGATCCATCAAC
For three sgRNA constructs	U6-29p-F: TTAATCCAAACTACTGCAGCCTGAC U6-1p-R: TATGCAAGTCTCACTCACACTCACG

Table 2.9: Primers used for sequencing of CRISPR/Cas9 constructs (Xing et al. 2014).

Primer	Sequence (5' - 3')
U6-26p-F	TGTCCCAGGATTAGAATGATTAGGC
U6-29p-F	TTAATCCAAACTACTGCAGCCTGAC

U6-29p-R	AGCCCTCTTCTTTCGATCCATCAAC
U6-29t-F	CGTGTTTCAGCTCTCATGATCCTTG

Table 2.10: Primers used for CRISPR/Cas9 construct detection in transgenic cell suspension cultures and tomato plants.

sgRNA sequences are in bold.

CRISPR/Cas9 construct	Sequence (5' - 3')
Solyc03g082450 (SYR2)	**82450gRNAF**: ATTG**AGGGATGCCACCAGTGAAA**G **U6-29p-R**: AGCCCTCTTCTTTCGATCCATCAAC
Solyc03g082470 (SYR1)	**82470gRNAF**: ATTG**CGGTAACTTGCCGGTCAAT**G **U6-29p-R**: AGCCCTCTTCTTTCGATCCATCAAC
Solyc01g109650 (LRRXIV)	**109650gRNAF**: TGATTG**CTGGGATTCGTTGCAAGAA**G **U6-29p-R**: AGCCCTCTTCTTTCGATCCATCAAC
Solyc07g063000 (PSKR2)	**63000gRNAF**: TAGTGATT**GGAACCTAATTGCTGTAAA**TG **U6-29p-R**: AGCCCTCTTCTTTCGATCCATCAAC
Soly09g083210 (LYK4)	**83210gRNAF**: TAGTGATTG**GTTACAAGAACACGTGAT**GG **U6-29p-R**: AGCCCTCTTCTTTCGATCCATCAAC
Solyc12g036330 (LRK10L1.2)	**36330gRNAF**: ATTG**CTGGTAAGCCACAGTAGGA**G **U6-29p-R**: AGCCCTCTTCTTTCGATCCATCAAC
Solyc02g070000 (GHR1)	**70000gRNAF**: TGATTGC**CAATCTCTGGTGGTAAAG**G **U6-29p-R**: AGCCCTCTTCTTTCGATCCATCAAC
Solyc03g123860 (PORK1)	**123860gRNAF**: AGTGATTG**GTACGACAGATTTAAACCC**G **U6-29p-R**: AGCCCTCTTCTTTCGATCCATCAAC

Table 2.11: Primers used for detection of mutations in transgenic cell suspension cultures and tomato plants.

CRISPR/Cas9 construct	Sequence (5' - 3')
Solyc03g082450 (SYR2)	**82450CRISPRF-136**: TCTGTCTCGCACTGCCAATGG **82450CRISPRR-737**: CCAAGATGAGCAGATGATGCAGAG
Solyc03g082470 (SYR1; 1st sgRNA)	**82470CRISPRF1-172**: AAGGGTGTTACCTGCTATTCAGAC **82470CRISPRR1-456**: GTAACCCAACTCAAGGTATACAAGC
Solyc03g082470 (SYR1; 2nd sgRNA)	**82470CRISPRF2-1071**: CACTTCTCTTGCTGAGATAAGTCTCG **82470CRISPRR2-1427**: AGTTTAGAAGGAATCTGTCCACTGAAG
Solyc01g109650 (LRRXIV)	**109650CRISPRF-328**: CATTCAGGAATAGAACCAGGAAGTAC **109650CRISPRR-72**: ACAGAGGGTTAGTTCTTCAACTGAG
Solyc07g063000 (PSKR2)	**63000CRISPRF-136**: GCTGGCAATCTAACAAATGG **63000CRISPRR-369**: CAAGTCCAAAGGCAATCCAC
Soly09g083210 (LYK4)	**83210CRISPRF-150**: GTTATACAGAGCTAGAGCACCAGAG **83210CRISPRR-716**: GATGATGTTGGCTGTGTAAGATTAGG
Solyc12g036330 (LRK10L1.2)	**36330CRISPRF-105**: GAGTTGTGGTGATGGAGTGAAC **33630CRISPRR-531**: ATCCTGACACGATTCGACTG
Solyc02g070000 (GHR1)	**70000CRISPRF-236**: TGCTTGTGAAACTCTCTATGGCT **70000CRISPRR-493**: GCCCAGAAAGAGAATTGTGACT
Solyc03g123860 (PORK1)	**123860CRISPRF-1428**: ATTCCTTCTCAATTAGGACAATGTCATAC **123860CRISPRR-2015**: CAGAAATGTTATTGCCAGCTATATCTAACTC

2.1.9. Vectors

Table 2.12: the vectors used in this study.

Vector	Description	Antibiotics Resistance
pCR2.1-TOPO®	Used for cloning of PCR products for sequencing and construct generation. Thermo Fischer Scientific (Waltham, United States).	Carbenicillin Kanamycin
pCBC-DT1T2	Used as template for amplifying the 1st and 2nd sgRNAs of two sgRNA CRISPR/Cas9 constructs and the 1st sgRNA of the three sgRNA construct. Addgene (Watertown, United States; Xing et al. 2014).	Chloramphenicol
pCBC-DT2T3	Used as template for amplifying the 2nd and 3rd sgRNA of the three sgRNA construct. Addgene (Watertown, United States; Xing et al. 2014).	Chloramphenicol
pKSE401	The binary vector, receiving the CRISPR/Cas9 constructs for transformation of tomato plants and *S. peruvianum* cell suspension cultures. Addgene (Watertown, United States; Xing et al. 2014).	Kanamycin
pET-21d (+)	The expression vector used to express MPK2 (WT and CA forms) and PLL5 PD in *E. coli* (Novagen, Darmstadt).	Carbenicillin
pET-21a (+)	The expression vector used to express GHR1 ID in *E. coli* (Novagen, Darmstadt).	Carbenicillin

2.1.10. Constructs

2.1.10.1. Design of CRISPR/Cas9 Constructs Targeting RLK Genes

Two-sgRNAs CRISPR/Cas9 constructs were designed to target mutations in all candidate RLKs except PORK1, which was mutated by a three-sgRNA CRISPR/Cas9 construct. The constructs were designed and cloned as described by Xing et al. (2014). Briefly, the gene loci of candidate RLKs were searched for target sequences using https://www.genome.arizona.edu/crispr/CRISPRsearch.html website. This website ranks target sequences according to their specificity and off-target potential. For all candidate RLKs, highly specific target sequences in the receptor ectodomain with the lowest off-target potential were selected. The sequence of the targets was inserted in the forward and reverse primers sequences (listed in Table 2.7). These primers were used to amplify the 1st target sequence along with the 1st sgRNA scaffold and terminator and the promoter of the 2nd sgRNA as well as the sequence of the 2nd target sequence using pCBC-DT1T2 vector (Addgene) as template (Figure 2.1). The PCR product was cloned by Golden Gate method into the Bsal site of pKSE401 (Addgene) binary vector, which already has the promoter sequence for the 1st sgRNA and the sgRNA scaffold of the 2nd target along with its terminator. This binary vector harbors a maize-codon optimized Cas9 gene controlled by a dual Cauliflower mosaic virus (CaMV) 35S promoter.

For the three-sgRNA CRISPR/Cas9 construct of PORK1 (performed by Latisha Lafleur), the sequence of the targets was inserted in the forward and reverse primers sequences (listed in Table 2.7). These primers were used to amplify the 1st target sequence along with the 1st sgRNA scaffold, its terminator and the promoter of the 2nd sgRNA using pCBC-DT1T2 vector (Addgene) as template (Figure 2.2). The primers for the 2nd and the 3rd targets were used to amplify the 2nd target sequence along with the 2nd sgRNA scaffold, its terminator, the 3rd sgRNA promoter and the 3rd target sequence using pCBC-DT2T3 (Addgene) as template. PCR products were cloned by Golden Gate method into the Bsal site of pKSE401 (Addgene) binary vector, which already has the promoter sequence for the 1st sgRNA and the sgRNA scaffold of the 3rd target along with its terminator.

After Golden Gate cloning (Engler et al. 2009), the ligated constructs were transformed in *E. coli* (DH10B). Primers in Table 2.8 were used for colony PCR to detect positive

E. coli clones. The correct assembly of the cloned binary vectors was confirmed by sequencing using the primers listed in Table 2.9. The purified binary vectors were transformed into *A. tumefaciens* GV3101 (Koncz and Schell 1986) with helper plasmid pSoup (Hellens et al. 2000).

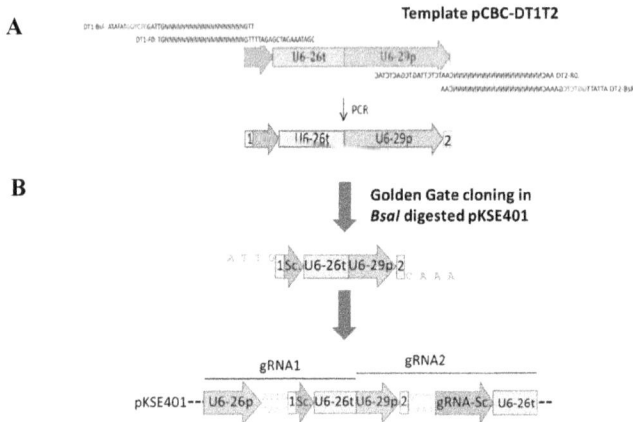

Figure 2.1: Diagram Summarizing the Cloning of the two-sgRNA CRISPR/Cas9 Constructs for RLKs.

(**A**) The 19N sequence in the primers was replaced with the target sequence chosen. An extra G at the 5' end of these 19N is already included in the designed primers. The RNA polymerase III-dependent U6 promoter requires this 5' G for transcription initiation (Zhang et al. 2017). These primers were used to amplify the cassette of the two-sgRNAs (1st gRNA scaffold with the 1st target sequence along with its terminator and the 2nd sgRNA promoter as well as the sequence of the 2nd target). (**B**) The PCR product was cloned by the Golden Gate method into the pKSE401 binary vector, which already has the sequence of the 1st sgRNA promoter and the 2nd sgRNA scaffold along with its terminator. The targets sequences are in yellow. The promoters are in light blue. The sgRNA scaffolds are in green. The terminators are in gray. The 5' and 3' overhangs of the PCR product after *Bsa*I digestion are shown.

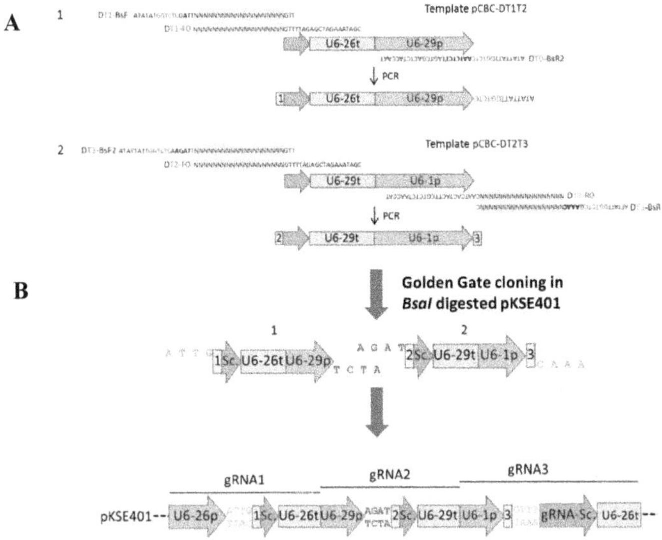

Figure 2.2: Diagram Summarizing the Cloning of the Three-sgRNA CRISPR/Cas9 Constructs for PORK1.

(A) The 19N sequences in the primers were replaced with the target sequences chosen. An extra G at the 5' end of these 19N is already included in the designed primers. The RNA polymerase III-dependent U6 promoter requires this 5' G for transcription initiation (Zhang et al. 2017). These primers were used to amplify the PCR product (1), which is the 1st sgRNA cassette (the 1st target sequence, the 1st sgRNA scaffold along with its terminator and the 2nd sgRNA promoter). PCR product (2) is the 2nd and 3rd sgRNA cassettes (the 2nd target sequence, the 2nd sgRNA scaffold along with its terminator, the 3rd sgRNA promoter and the 3rd target sequence). (B) The PCR products were cloned by Golden Gate method into the pKSE401 binary vector, which already has the sequence of the 1st gRNA promoter and the 3rd gRNA scaffold along with its terminator. The targets sequences are in yellow. The promoters are in light blue. The gRNA scaffolds are in green. The terminators are in gray. The 5' and 3' overhangs of the PCR product after BsaI digestion are shown.

2.1.10.2. MPK2, PLL5 Phosphatase Domain and GHR1 Intracellular Domain Expression Constructs

Primers listed in Table 2.5 were used to amplify the coding sequence of MPK2 in wild-type (WT) and constitutively active (CA) forms, PLL5 PD and GHR1 ID from tomato leaf cDNA. The amplified PCR products were cloned into pCR2.1-TOPO® (Thermo Fisher Scientific) and transformed in E. coli (DH10B). After confirming the correct sequence by sequencing, the coding sequences were digested and ligated in pET-21a (+) or pET-21d (+) (Figure 2.3). The ligated constructs were transformed into *E. coli* (DH10B) for amplification. Finally, the constructs were transformed into *E. coli* BL21 RIL (Stratagene) for expression.

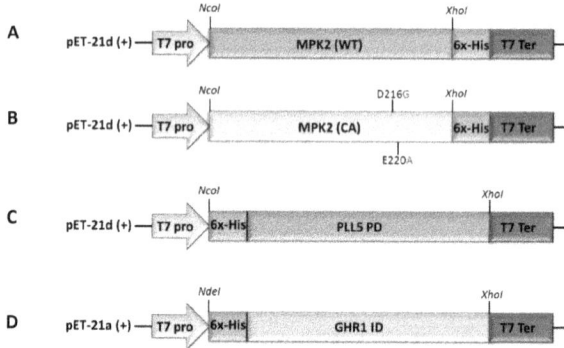

Figure 2.3: MPK2 (WT and CA forms), PLL5 PD and GHR1 ID Constructs Expressed in *E. coli*. A diagram showing the expression cassettes of the WT (**A**) and the CA (**B**) forms of MPK2 and the PLL5 PD (**C**), which were cloned in the *NcoI/XhoI* sites of pET-21d (+), and the GHR1 ID (**D**) that was cloned in the *NdeI/XhoI* sites of pET-21a (+). The constructs were transformed in *E. coli* BL21 RIL (Stratagene) for expression of the recombinant proteins.

2.2.   Organisms

2.2.1. Bacterial Strains

Table 2.13: Bacterial strains used in this work.

Organism/strain	Genotype	Antibiotic Resistance
E. coli DH10B™ (Invitrogen)	F-mcrA Δ(mrr-hsdRMS-mcrBC) Φ80lacZΔM15 ΔlacX74 recA1 endA1 araD139Δ(ara-leu)7697 galU galK λ-rpsL nupG	-
E. coli BL21 RIL (Stratagene)	*E. coli* B F⁻ ompT hsdS(rB⁻ mB⁻) dcm⁺ Tetr gal endA Hte [argU ileY leuW Camʳ].	Chloramphenicol

	argU (AGA, AGG), ileY (AUA), leuW (CUA)	
A. tumefaciens GV3101 (pMP90; Koncz and Schell 1986)	C58 (rifR) Ti pMP90 (pTiC58DT-DNA) (gentR) Nopaline (pSoup-tetR)	Rifampicin and Gentamicin

2.2.2. Bacterial Stocks

Table 2.14: Bacterial stocks produced during this work.

Construct	Organism	Serial Numbers
SYR2 CRISPR/Cas9 construct	*E. coli* (DH10B)	4578
	A. tumefaciens (GV3101+pSoup)	4579
SYR1 CRISPR/Cas9 construct	*E. coli* (DH10B)	4580
	A. tumefaciens (GV3101+pSoup)	4581
LRRXIV CRISPR/Cas9 construct	*E. coli* (DH10B)	4582
	A. tumefaciens (GV3101+pSoup)	4583
LYK4 CRISPR/Cas9 construct	*E. coli* (DH10B)	4584
	A. tumefaciens (GV3101+pSoup)	4585
PSKR2 CRISPR/Cas9 construct	*E. coli* (DH10B)	4586
	A. tumefaciens (GV3101+pSoup)	4587
LRK10L1.2 CRISPR/Cas9 construct	*E. coli* (DH10B)	4588
	A. tumefaciens (GV3101+pSoup)	4589

GHR1 CRISPR/Cas9 construct	E. coli (DH10B)	4774
	A. tumefaciens (GV3101+pSoup)	4776
PORK1 CRISPR/Cas9 construct	E. coli (DH10B)	5139
	A. tumefaciens (GV3101+pSoup)	5140
6XHis-PLL5-PD	E. coli (DH10B)	5007
	E. coli (BL21 RIL)	5008
6XHis-GHR1-ID	E. coli (DH10B)	5011
	E. coli (BL21 RIL)	5012
MPK2WT-6XHis	E. coli (DH10B)	5017
	E. coli (BL21 RIL)	5018
MPK2CA-6XHis	E. coli (DH10B)	5019
	E. coli (BL21 RIL)	5020

2.2.3. Solanum lycopersicum

All RLK CRISPR/Cas9 constructs were transformed into tomato (*S. lycopersicum*) cv. UC82B plants (performed by Dagmar Repper and Latisha Lafleur).

Tomato plants cv. UC82B overexpressing Prosystemin and the double-mutated Prosystemin were kindly provided by Dr. Konrad Dreizler (Dreizler 2018).

Wild-type UC82B plants were used as controls for wound experiments.

2.2.4. *Solanum peruvianum*

The wild-type cell suspension culture was created from wild-type tomato callus (*Solanum peruvianum*) and kindly provided by Prof. Georg Felix, University of Tübingen.

The cells of *S. peruvianum* were transformed with the RLK CRISPR/Cas9 constructs for the generation of RLK loss-of-function cell suspensions as described in section 2.9.

The cultures were grown in 70 ml Nover medium (section 2.1.6.2) in a shaker (Infors HT Multitron Pro, Switzerland) at 120 rpm and 26°C. The transgenic cell suspension cultures were supplemented with 75mg/l Kanamycin. About 1/10 of the culture volume was transferred to fresh medium weekly (performed by Ursula Glück-Behrens).

2.3. Molecular Biology Methods

2.3.1. Isolation of Genomic DNA from Tomato Leaves or *S. peruvianum* Cells

Tomato leaves or *S. peruvianum* cells frozen in liquid N_2 were ground into a fine powder using a tissue lyser (TissueLyser LT; Qiagen) and homogenized in 500 µl DEX buffer (220 mM Tris/HCl pH 8.0, 222 mM EDTA pH 8.0, 800 mM NaCl, 140 mM sorbitol, 0.8% (w/v) cetyltrimethylammonium bromide, 1% (w/v) sarcosine, 2% (v/v) β-mercaptoethanol). The homogenate was incubated for 30 min at 65°C, and extracted with 1 volume of chloroform. After centrifugation at 16,000 x g for 10 min, the clear aqueous phase was precipitated with 2.5 volumes of EtOH. The DNA pellet collected by centrifugation was washed with 70% (v/v) EtOH, dried and dissolved in 30-40 µl sterile ddH$_2$O. The isolated DNA was stored at -20°C.

2.3.2. Standard-Polymerase Chain Reaction (PCR)

Standard PCR was performed in a total volume of 25 µl in a T100™ Thermo Cycler (BioRad). The following components were used for each reaction.

Reaction Mix:

5 µl of 5x PCR Buffer (15 mM $MgCl_2$, 100 mM $(NH_4)_2SO_4$, 0.08% (v/v) Triton X-100, 20% (v/v) DMSO, 250 mM KCl, 50 mM Tris/HCl pH 8.3), 0.5 µl dNTPs (10 mM; Bioline, Luckenwalde), 1 µl of *Taq* DNA Polymerase, 0.5 µl of each of the forward (10 µM) and the reverse primers (10 µM), 15.5 µl ddH₂O and 2 µl template DNA.

The amplification reaction was carried out under the following conditions:

2 min at 95°C   1 cycle

30 sec at 95°C
30 sec at 60°C   35 cycles
1 min at 72°C

10 min 72°C   1 cycle

∞ at 10°C   1 cycle

For the amplification of MPK2, PLL5 PD and GHR1 ID coding sequences, 0.5 µl of Advantage® 2 PCR enzyme (Takara Bio, United States) along with its buffer were used instead of the *Taq* polymerase and the 5x PCR buffer. For these reactions the PCR program was 20-24 cycles.

2.3.3. DNA Agarose Gel Electrophoresis

DNA was separated using 1-3% (w/v) TAE agarose gels (50×TAE buffer: 242 g Tris-base, 57.1 ml of acetic acid, 100 ml of 0.5 M EDTA pH 8.0, 14 µg/l ethidium bromide). For the preparation of a gel, 1x TAE buffer with 1-3 g agarose LE (Genaxxon, Ulm) were mixed and 1 µl (10 mg/ml) ethidium bromide/100 ml gel volume was added. Before loading, DNA was mixed with 1-2 µl of 10x DNA loading buffer (0.1% (w/v) Xylene cyanole, 50% (v/v) glycerol, 50% (v/v) 1x TAE buffer [40 mM Tris, 20 mM acetic acid, 1 mM EDTA pH 8.9]). The size marker was 4-5 µl of a DNA ladder

(GeneRuler™ 1 kb or 100 bp, Thermo Fisher Scientific). Electrophoresis was carried out at 8-12 volts/cm of gel. After separation of the DNA, the gel was visualized by GelDoc™ EZ Imager (BioRad, Munich) and photo-documented with Image Lab 4.0 (BioRad, Hercules, USA).

2.3.4. DNA Native Polyacrylamide Gel Electrophoresis (PAGE)

For genotyping CRISPR/Cas9-mediated indel mutations in transgenic tomato plants and *S. peruvianum* cells, DNA native PAGE was used as described by Zhu et al. (2014). This method helped to detect heterozygous and multiallelic mutated plants or cells and differentiate them from wild-type and homozygous mutated plants or cells. After denaturation and annealing of PCR products, homoduplexes of identical DNA strands are formed as well as heteroduplexes of DNA strands from different alleles (Figure 2.4 A). In native PAGE, heteroduplex DNA migrates slower than homoduplex due to the formed open angle between matched and unmatched genomic regions, which allow their identification based on their mobility rate (Figure 2.4 B).

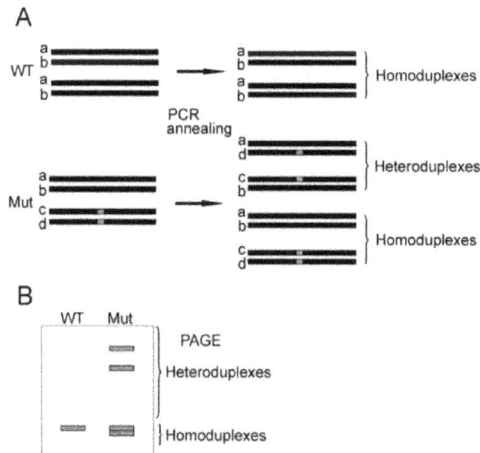

Figure 2.4: A Schematic Overview of DNA Native PAGE Method Used to Genotype CRISPR/Cas9-mediated Indel Mutations in Transgenic Tomato Plants and *S. peruvianum* Cells. (A) During PCR denaturation and annealing, homoduplexes and heteroduplexes form. Dark green bars represent four DNA strands (a–d) in heterozygous cells harboring a mutated allele (bars with orange boxes) and a wild-type allele. After denaturation and annealing, two types of homoduplex DNA and two

types of heteroduplex DNA are formed. (**B**) Homo- and heteroduplexes can be identified by running them on 10% native PAGE. Due to formation of an open angle between matched and unmatched genomic regions, heteroduplex DNA migrates slower and can be easily distinguished from homoduplex DNA. (Adapted from Zhu et al. 2014).

For this method 10% native PAGE was prepared by mixing: 5 ml of acrylamide stock solution (Rotiphorese® Gel40, Roth), 4 ml of 5x TBE buffer (0.45 M Tris-Borate, 0.01M EDTA, pH 8.3), 11 ml ddH$_2$O, 200 µl of 10% (w/v) Ammonium persulfate (APS) and 20 µl TEMED. The mixture was poured in a gel cassette (1.5 mm) and left to polymerize for about 30 min.

About 15 µl of PCR products spanning mutation target sites of RLKs (performed using the primers in Table 2.11) mixed with 2µl of 10x DNA loading buffer were loaded onto the polymerized native PAG. After 2-3 hours of electrophoresis at 120 V at 4°C, polyacrylamide gel was immersed in 0.5% (v/v) ethidium bromide solution (in 1x TBE buffer) for 10 minutes before visualization using GelDoc™ EZ Imager (BioRad, Munich) and documented using Image Lab 4.0 (BioRad, Hercules, United States).

The PCR products that run on native PAGE as homoduplexes were sent to sequencing to find out if they have wild-type or mutated sequences.

2.3.5. Elution of DNA Fragments from Agarose Gel

The DNA fragments were eluted from the agarose gels using QIAquick Gel Extraction Kit (Qiagen, Hilden) according to the manufacturer to instructions.

2.3.6. Restriction Digestion

For analytical purposes, 3 µl of plasmid DNA or 7 µl of PCR products (about 100 ng/µl), 2 µl of 10x enzyme buffer (specific for the particular enzyme) and 0.3 µl (3 U) of restriction enzyme were mixed in a 0.2 ml reaction tube and supplemented with ddH$_2$O up to 20 µl final volume. If the DNA fragments were to be used for further cloning steps, the reactions were scaled up five times.

After an incubation period of 1.5-2 hours at 37°C, the reaction was analyzed by gel electrophoresis (section 2.3.3).

2.3.7. Dephosphorylation of DNA Fragments

At the end of the restriction digestion (2.3.6) of vectors, 1 μl of thermosensitive alkaline phosphatase (FastAP, Thermo Fisher Scientific) was added to the reaction tubes. This served to remove the 5' phosphate residue to prevent self-ligation of the vector. The reaction was stopped by adding 5 mM EDTA with 10 min incubation at 75°C.

2.3.8. DNA Ligation

The ligation mixture was composed of: insert and vector in a ratio of 3:1, 1 μl of ligation buffer (10x), 1 μl of ATP (10 mM) and 1 μl of T4 DNA ligase (Fermentas, St. Leon-Rot). The entire reaction mixture was made up to 10 μl with ddH$_2$O. The ligation was performed under the following conditions:

8 hrs	at 16°C
1 hr	at 30°C
5 min	at 70°C
∞	at 10°C

2.3.9. Golden Gate Cloning of CRISPR/Cas9 Constructs

For constructing the CRISPR/Cas9 constructs the following reaction was setup by adding:

Two hundred and fifty fmol of the purified PCR fragments for two-sgRNAs or three-sgRNAs CRISPR/Cas9 constructs as illustrated in Figures 2.1A and 2.2A respectively, 15 fmol of pKSE401 vector (about 100 ng/μl), 1.5 μl of 10x T4 DNA Ligase Buffer (New England Biolabs), 1.5 μl of 100 μg/ml BSA, 1 μl *BsaI* (New England Biolabs), 1 μl of T4 DNA Ligase (2,000,000 Unit/ml, New England Biolabs) and ddH$_2$O up to 15 μl.

The reaction was performed under the following conditions:

5 hrs	at 37°C
5 min	at 50°C
10 min	at 80°C
∞	at 10°C

2.3.10. TOPO® Cloning

PCR products generated using the primers listed in Table 2.5 for MPK2 CA, MPK2 WT, PLL5 PD and GHR1 ID expression constructs, and in Table 2.11 for the genotyping of mutated cell suspension cultures were cloned into pCR2.1-TOPO® (Thermo Fisher Scientific) vector according to manufacturer instructions.

2.3.11. Transformation of Electrocompetent *E. coli* and *A. tumefaciens*

The electrocompetent cells (40 μl) were thawed slowly on ice, 1 μl of plasmid DNA was added, incubated on ice for 1 min and then transferred to cooled electroporation cuvettes with 0.1 cm electrode spacing. Electroporation (MicroPulser™, BioRad) was performed at 2.5 kV (*E. coli*) or 2.2 kV (*A. tumefaciens*). The cells were then immediately taken up in 1 ml SOC medium into a 15 ml culture tube and incubated for one hour at 37°C for *E. coli*, and three hours at 28°C for *A. tumefaciens* in an orbital shaker (225 rpm). Subsequently, 50 μl of the bacterial suspension were plated on LB plates with appropriate antibiotics. The remaining bacterial suspension was centrifuged for 1 min at 16,000 x g, the supernatant discarded to about 50 μl, cells pellet resuspended and plated as well.

2.3.12. Transformation of Chemically Competent *E. coli*

Fifty µl of chemically competent *E. coli* cells were thawed on ice and 1 µl of plasmid DNA or ligation product was added, mixed gently and incubated on ice for 30 min. The mixture was heat shocked at 42°C for 2 min and then cooled on ice for 3 minutes. Then 1 ml of SOC medium was added to the bacteria, which were incubated for 1 hr at 37°C with shaking at 225 rpm. Thereafter, the cells were spread on LB medium plates with the appropriate antibiotics. If blue-white selection was performed, 20 µl of 100 mM Isopropyl β-D-1-thiogalactopyranoside (IPTG) and 40 µl of 20 mg/ml 5-bromo-4-chloro-3-indoyl-β-D-Galactopyranoside (X-Gal) were spread out on the LB medium before plating of the bacteria.

2.3.13. Colony PCR

A single colony of *E. coli* was picked with a pipette tip and resuspended in 5 µl of sterile ddH$_2$O.

After addition of 22 µl of PCR reaction mixture (section 2.3.2), the bacterial cells were first disrupted for 5 min at 95°C. Then a standard PCR was carried out (section 2.3.2). The clones which had a positive PCR product were incubated overnight at 37°C in 3 ml LB medium plus antibiotics (section 2.3.14). They were either stored as a permanent culture at -80°C in 15% (v/v) glycerol or used for plasmid isolation (section 2.3.15).

2.3.14. Bacterial Cultures

Transformed *E. coli* or *A. tumefaciens* were cultured on solid LB medium with the appropriate antibiotics overnight at 37°C and 2-3 days at 28°C, respectively. For plasmid isolation or protein expression, a single colony was transferred to 3 ml of LB with appropriate antibiotics. For *E. coli*, the liquid cultures were cultured at 37°C with shaking (220 rpm) for 24 hrs, for *A. tumefaciens* at 28°C for 48-72 hrs.

2.3.15. Plasmid DNA Isolation (Mini-Preparation) from *E. coli*

A single bacterial colony was incubated overnight in 2 ml of LB medium with the appropriate antibiotics at 37°C and 225 rpm in an orbital shaker. The next day, 1.5 ml of this culture were transferred to a reaction tube and the bacteria were sedimented for 2 minutes at 16,000 × g. The supernatant was discarded, and the resulting sediment resuspended in 100 μl mini-prep. solution I (50 ml ddH$_2$O, 25 mM Tris/HCl, pH 7.5, 50 mM glucose, 10 mM EDTA pH 8). Cells were lysed by adding 150 μl mini-prep. solution II (0.2 N NaOH, 1% (w/v) SDS), mixed by inversion and incubation on ice for 5 min. Then, 150 μl mini-prep. solution III (60 ml 5 M potassium acetate, 11.5 ml glacial acetic acid, 28.5 ml ddH$_2$O) were added, mixed by short inversion and incubated 5 min on ice. Upon centrifugation for 10 min at 4°C and 16,000 x g, the obtained supernatant with the plasmid DNA was transferred into a new reaction tube. This was followed by addition of twice the volume of 99% (v/v) EtOH, 15 min incubation on ice and 15 min centrifugation at 4°C and 16,000 x g. The resulting DNA precipitate containing the plasmid DNA was re-suspended, after removal of the ethanol, in 200 μl of TE buffer (10 mM Tris/HCl pH 8.0, 1 mM EDTA pH 8). The plasmid solution was incubated for 30 min at 37°C with 50 U pancreatic RNase (10 mg/ml in TE) to remove the RNA. The mixture was then extracted with 1 volume of phenol/chloroform (1:1 v/v), centrifuged for 2 min at 16,000 x g and the supernatant transferred to a new reaction tube containing 1 volume of chloroform and mixed by vigorous shaking. After centrifugation as above, the supernatant was transferred to another reaction tube and the plasmid DNA was precipitated with 1/10 volume of 3 M sodium acetate (pH 5.2) and 2.5 volumes of 99% (v/v) EtOH and incubated on ice for 30 min. This was followed by a centrifugation step of 15 min at 16,000 x g at 4°C. The precipitate was washed in 70% (v/v) EtOH, dried and dissolved in 20-30 μl ddH$_2$O. The isolated plasmids were stored at -20°C.

2.3.16. DNA Sequencing

The plasmid samples and PCR products were prepared for sequencing according to company specifications (Macrogen Europe, Amsterdam, The Netherlands). The results were analyzed using CLC Main Workbench v. 8.0 (Qiagen Bioinformatics).

2.3.17. Isolation of RNA from Tomato Leaves

Tomato leaves were placed in a mortar and with a pestle ground into a fine powder in liquid N_2. Subsequently, 50-100 mg of the tissue was extracted with Trizol (peq GOLD TriFast™, VWR, Darmstadt) and the samples incubated for 10 min at RT in order to ensure the dissociation of nucleotide complexes. After 10 min centrifugation at 4°C and 16,000 x g, the RNA-extract was transferred to a new reaction tube. To ensure complete removal of proteins, the samples were extracted four times with an equal volume of chloroform and the aqueous phase was recovered by centrifugation for 5 min at 4°C and 16,000 x g. Then same volume of isopropanol (v/v) was added and the RNA was precipitated for 15 minutes on ice. RNA was sedimented by centrifugation for 10 min at 4°C and 16,000 x g. To remove salts and other contaminants, the RNA pellet was then washed twice with 75% EtOH (v/v) for 5 min at 4°C and 16,000 x g. After complete removal of the supernatant, the RNA sediment was dried for 5-15 minutes at room temperature then dissolved in 20 µl ddH₂O. Subsequently the RNA was stored at -80°C. The RNA concentration was measured at 260 nm using Tecan Spark® (Männedorf, Switzerland).

2.3.18. Reverse Transcriptase Reaction (cDNA Synthesis)

For the reverse transcriptase reaction, the following components were mixed in 0.2 ml reaction tubes:

1 µl of random hexamer primers (100 µM, Thermo Fisher Scientific), 1 µl of RevertAid reverse transcriptase (200 U/µl, Thermo Fisher Scientific), 4 µl of 5x reaction buffer (Thermo Fisher Scientific), 2 µl of dNTPs (10 mM, Bioline GmbH, Luckenwalde), 2 µg RNA and ddH₂O up to 20 µl final volume. The reaction was performed in a thermocycler (T100™ Thermo Cycler, BioRad) using the following parameters: 10 min at 25°C, 60 min at 42°C, and 10 min at 70°C. The resulting cDNA was then stored at -20°C for further analytical purposes. For real-time PCR, the cDNA samples were diluted 1:10 in ddH₂O.

2.3.19. Quantitative Real-time PCR (qRT-PCR)

Real-time PCR was used for the quantification of RLK and *PI-II* gene expression levels after wounding using the primers listed in Table 2.6. SYBR Green (10,000x concentrate in DMSO, Cambrex, Rockland, United States) was used for relative quantification, where the expression of the gene of interest is related to that of housekeeping genes (elongation factor 1α (*Sl*EF1α), *Sl*Actin and *Sl*Ubiquitin for *PI-II*, only *Sl*EF1α for RLKs). The PCR reaction was carried out in 96-well plates in a CFX96 Real-Time PCR System (BioRad).

The reactions were prepared as follows: 4 μl 5x PCR Buffer (15 mM MgCl$_2$, 100 mM (NH$_4$)$_2$SO$_4$, 0.08% (v/v) Triton X-100, 20% (v/v) DMSO, 250 mM KCl, 50 mM Tris/HCl pH 8.3), 0.5 μl dNTPs (10 mM; Bioline, Luckenwalde), 1 μl of *Taq* DNA Polymerase, 2 μl Forward primer/Reverse primer (1 μM) mixed 1:1, 2 μl SYBR Green (1/4000), 5.5 μl ddH$_2$O and 5 μl template cDNA (diluted 1:10).

The reaction was carried out under the following conditions:

95°C for 5 min   1 cycle

95°C for 30 sec ⎤
60°C for 40 sec ⎬ 40 cycles
72°C for 40 sec ⎦

Detection of fluorescence increase was measured after the extension step. To ensure uniformity of PCR products, a melting curve from 65°C to 95°C was performed afterwards. The results were evaluated with the Bio-Rad CFX Manager 3.1 (BioRad).

2.4. Protein Biochemical Methods

2.4.1. Expression of Recombinant 6xHis-tagged MPK2 (WT and CA forms), PLL5 PD, and GHR1 ID in *E. coli*

A single colony of the *E. coli* expression strain BL21 RIL transformed with MPK2 (WT or CA), PLL5 PD or GHR1 ID was used to inoculate a preculture in 3 ml LB medium including the appropriate antibiotics and incubated overnight at 37°C and 230 rpm in an orbital shaker. The next day, the main culture was inoculated with 0.5% (v/v) of the preculture and the bacteria were grown to an $OD_{600} = 0.8\text{-}1.0$. To induce protein expression, IPTG was added to a final concentration of 1 mM and incubated for at least 4 hrs at 30°C and 225 rpm. The main culture was then centrifuged at 6084 x g and 4°C for 20 min, the supernatant discarded, and the cell sediment stored at -70°C.

2.4.2. Affinity Purification of Recombinant 6xHis-tagged MPK2 (WT and CA), PLL5 PD and GHR1 ID on Ni^{2+}-NTA Agarose

The bacterial sediments (section 2.4.1) were thawed on ice and resuspended in 5ml/g of sediment BugBuster™ protein extraction reagent (Merck, Darmstadt, Germany). One spatula tip of DNase I (Thermo Fisher Scientific) was added, and the solution was incubated for 20 min at room temperature with 400 rpm shaking. By the end of incubation, 100 µl sample was removed into a 1.5 ml reaction tube, which was centrifuged for 10 min at 14,000 rpm at 4°C. The supernatant was collected to a new tube (fraction 2, raw extract) and the pellet was kept (fraction 1, insoluble fraction). The rest of the bacterial lysate was centrifuged for 20 min at 20483 x g and 4°C, the supernatant was transferred to a new reaction tube and its volume was determined.

For the purification of the 6xHis-tagged recombinant proteins, Ni^{2+}-NTA agarose from Qiagen was used. About 1.5 ml of nickel agarose slurry was added to a column, and the excess liquid was drained. The beads were washed with 4 ml of column washing buffer (50 mM Sodium Phosphate Buffer pH 7), equilibrated with 6 ml of binding buffer (50 mM sodium phosphate buffer pH 7, 300 mM NaCl, 20 mM imidazole, 4 mM benzamidine). This step was repeated three times. The protein lysate was added to the equilibrated Ni^{2+}-NTA agarose, the agarose resuspended therein, the column top and bottom closed and incubated in an end-over-end shaker at 4°C. After one hour, the flow-through was drained and collected. The column was then washed four times with 4 ml of binding buffer. Wash solutions 1-4 were collected separately. For the elution of the bound proteins, 0.75 ml of elution buffer (50 mM sodium phosphate buffer pH 7, 300 mM NaCl, 250 mM imidazole, 4 mM benzamidine) was added to the column. The elution was repeated four times and eluates 1-4 were collected separately. For column regeneration and storage, the agarose beads were washed with 8 ml of elution buffer, then 8 ml of column washing buffer (50 mM sodium phosphate buffer pH 7), and at the end with 8 ml of 30% (v/v) EtOH. For preservation of the column, the agarose beads were stored in the column in 2 ml of 30% (v/v) EtOH at 4°C.

Like for the insoluble and raw extract fractions, samples from the flow through, washes 1-4 and eluates 1-4 (fractions 3-9) were taken and placed in ice. Samples of these fractions were then mixed with 4×SDS Buffer (0.2 M Tris/HCl, 0.4 M DTT, 8% (w/v) SDS, 0.4% (w/v) bromophenol blue (Serva, Heidelberg), 40% (v/v) glycerol), denatured for 5 minutes at 95°C, chilled on ice, centrifuged briefly and stored at -20°C until SDS-PAGE (section 2.4.5).

2.4.3. Dialysis of the Purified Recombinant 6xHis-tagged MPK2 (WT and CA), PLL5 PD, and GHR1 ID

A few cm of dialysis tube (12.0S, 20 µM wall thickness, ZelluTrans, Roth, molecular exclusion limit 12 kDa) were soaked in water for 15 min, transferred to 400 ml of 10 mM sodium bicarbonate and incubated at 80°C for 10 min with stirring. Subsequently, the tubes were incubated in ddH$_2$O for further 10 min at 80°C with stirring. Eluate 2 of the purified recombinant proteins were filled into the prepared tubes, closed with two clamps and dialyzed 3 times for 6 hrs against 500 times their volume of the storage buffer (25 mM HEPES pH 7.5, 100 mM NaCl, 1mM DTT, 0.1 mg/ml BSA) at 4°C. Finally, the protein solutions were transferred to 1.5 ml tubes and stored at 4°C.

2.4.4. Protein Concentration Measurement

Protein concentration in microsomal fractions (section 2.4.9) was measured using Bradford assay (Bradford 1976). Two μl sample were mixed with 248 μl of 1x Bradford reagent (Roti®-Quant, Roth) in a 96-well microplate, incubated for 5-10 min, and the absorbance was measured at 595 nm using Tecan Spark®. Protein concentration was then determined for three technical replicates using a Bovine Serum Albumin (BSA) standard curve with a range of 1-15 μg/ml.

The concentration of purified recombinant 6xHis-tagged proteins (sections 2.4.2 and 2.4.3) and synthetic peptides (Table 2.4) was measured using DC™ Protein Assay kit (BioRad), according to manufacturer's instructions. The absorbance was read at 750 nm using Tecan Spark®. The protein concentration was then determined for three technical replicates using a BSA standard curve with a range of 1-15 μg/ml.

2.4.5. SDS-PAGE

Protein samples collected in section 2.4.2 were loaded on a denaturing SDS polyacrylamide gel prepared as follows: separating gel: 375 mM Tris/HCl pH 8, 0.1% (w/v) SDS, 10% (v/v) acrylamide solution (Rotiphorese® Gel40, Roth), 0.03% (w/v) APS, 0,075% (v/v) TEMED; stacking gel: 125 mM Tris/HCl pH 6.8, 0.1% (w/v) SDS, 5% (v/v) acrylamide solution (Rotiphorese®Gel40, Roth), 0.03% (w/v) APS, 0.18% (v/v) TEMED. The samples were run on the gel for about 1.5 hr at 120 V using a running buffer of 50 mM Tris/HCl pH 8.3, 384 mM glycine, 0.1% (w/v) SDS. Five μl of BlueStar Prestained Protein Marker (Nippon Genetics Europe, Düren) was loaded on the gel to estimate protein mass.

At the end of the run, gels were stained for 30 min in Coomassie solution (0.25% (w/v) Brilliant Blue R 250 (Serva, Heidelberg), 45% (v/v) methanol, 10% (v/v) glacial acetic acid and 44.75% ddH$_2$O). Then it was de-stained overnight in 30% (v/v) methanol, 10% (v/v) glacial acetic acid until the protein bands were clearly visible. For imaging, the gels were photographed with the Gel Doc™ EZ Imager (BioRad).

2.4.6. Western Blot

Proteins separated by SDS-PAGE (section 2.4.5) were transferred onto Amersham™ Protran® Premium nitrocellulose membrane (Sigma-Aldrich) using Trans-Blot® SD Semi-Dry Transfer Cell (Biorad). For this purpose, the membrane (9 cm x 5.5 cm) was washed with ddH$_2$O for 1 min before incubation in anode buffer (48 mM Tris-HCl, 39 mM glycine, 1.3 mM SDS, 20% (v/v) methanol) for 10 min. To assemble the blotting sandwich, six blotting papers (Whatman 3 mm chromatography paper, GE Healthcare, Freiburg) soaked in anode buffer were placed onto the anode plate followed by the nitrocellulose membrane and the polyacrylamide gel, avoiding air bubbles. Six blotting papers soaked in cathode buffer (48 mM Tris-HCl, 39 mM glycine, 1.3 mM SDS) were placed on top and then covered by the cathode plate. The transfer was performed at 250 mA per blot and 25 V for 1.5 hr.

The membrane was incubated in a blocking solution (1% (w/v) BSA in TBS-T (137 mM Nacl, 27 mM KCl, 25 mM Tris pH 7.4, 0.1% (v/v) Tween20)) for at least 1 h at room temperature. The anti His-tag primary antibody (Dianova, Hamburg) was added to the blocking solution (1:7500) and incubated overnight at 4°C with shaking, followed by three washing steps with TBS-T for 10 min each. Then, the secondary antibody (Rabbit Anti-Mouse IgG HRP-conjugate, Merck) diluted in blocking solution (1:10,000) was applied onto the membrane for 2 hrs at room temperature with shaking. After that, three washing steps with TBS-T were performed as described before. Immunolabelled proteins were visualized using the SuperSignal R West Dura Extended Duration Substrate kit (Thermo Fisher Scientific), following the manufacturer's instructions. Chemiluminescence was detected after 10 min using an Odyssey® Fc imaging system (LI-COR Biosciences, United States).

2.4.7. *In vitro* Kinase Assay

To make sure that the MPK2 WT and CA forms as well as GHR1 ID were catalytically active, 2 µg of purified proteins were incubated with 6 µg of myelin-basic protein (MBP, Sigma-Aldrich) for 1 hr in 50 µl of reaction buffer (20 mM HEPES pH 7.0, 10 mM MgCl$_2$, 2 mM DTT and 0.1 mg/ml BSA) with 1 µM ATP at 30°C. Kinase activity was measured as ATP consumption by addition of 50 µl of Kinase-Glo® Plus Reagents

(Promega; Fitchburg, WI). After 10 min incubation at room temperature, the luminescence (560 nm) was detected using Tecan Spark®.

Recombinant MPK2 CA and WT (2 μg) were incubated with peptide substrate (LHA1T955: GLDIETIQQSYTV; amount as indicated in Figure 3.15) using the same conditions as for MBP but the reaction was left overnight at 30°C. Kinase-Glo® Plus Reagents was used to detect ATP consumption as described above for MBP. An ATP standard curve with a range of 0.25 – 2 μM was used for quatification.

To confirm phosphorylation of LHA1T955 by mass spectrometry, the reaction was performed as described above with 20 μg of peptide substrate. The reaction was stopped by addition of 10% (v/v) trifluoroacetic acid (TFA) until pH ≤ 3 and reaction products were purified over C18-Stage tips (section 2.4.13) prior to mass spectrometric analysis (section 2.4.14).

2.4.8. *In vitro* Phosphatase Assay

Recombinant PLL5 PD (1 μg) was incubated with the phosphorylated peptide substrate (LHA1T955(ph): GLDIETIQQSYT(ph)V; amount as indicated in Figure 3.13) in 50 μl of reaction buffer (20 mM Tris-HCl pH 7.5, 5 mM $MgCl_2$, 1 mM EGTA, 0.02% (v/v) β-mercaptoethanol and 0.1 mg/ml BSA) at room temperature for 30 minutes. Released phosphate was detected as described by He and Honeycutt (2005) by addition of 150 μl of 8% (w/v) ascorbic acid, 2.1% (v/v) HCl and 0.7% (w/v) ammonium molybdate. After 5 min incubation at room temperature the reaction was stopped by addition of 150 μl of 2% (w/v) tri-sodium citrate, 2% (v/v) acetic acid. After incubation for 15 min at room temperature the reduced phosphomolybdenum complex was quantified at 640 nm using Tecan Spark® based on a standard curve of 0.25 – 2 mM K_2HPO_4.

To confirm dephosphorylation of LHA1T955(ph) by mass spectrometry, the reaction was performed as above with 200 μg of phosphorylated peptide substrate. The reaction was stopped by addition of TFA to reach pH ≤ 3 and reaction products were purified over C18-Stage tips (section 2.4.13) prior to mass spectrometric analysis (section 2.4.14).

2.4.9. Extraction of the Microsomal Membrane Fraction from *S. peruvianum* Cells

Microsomal membranes from *S. peruvianum* cells were extracted according to Wu et al. (2017) with minor modifications.

About 1 g of cells was homogenized in 10 ml extraction buffer (330 mM sucrose, 100 mM KCl, 1 mM EDTA, 50 mM Tris-MES pH 7.5 and fresh 5 mM DTT) in the presence of 1% (v/v) proteinase inhibitor mixture (Serva) and 0.5% (v/v) of each phosphatase inhibitor cocktails 2 and 3 (Sigma-Aldrich) in a Dounce Homogenizer on ice. At least 200 strokes were performed slowly to avoid formation of air bubbles. The homogenate was filtered through four layers of miracloth (VWR, Radnor, United States) and centrifuged for 15 min at 7500 × g at 4°C. The supernatant was collected and centrifuged for 75 min at 48,000 × g at 4°C. The pellet was washed with 5 ml of 100 mM Na_2CO_3 then again was centrifuged for 75 min at 48,000 x g at 4°C. The microsomal membrane pellets used for LC-MS/MS (section 2.4.14) were resuspended in 100 μl UTU buffer (6 M urea, 2 M thiourea and 10 mM Tris-HCl pH 8). Microsomal membrane pellets used for ATPase activity assay (section 2.4.10) were resuspended in 100 μl of resuspension buffer (330 mM sucrose, 25 mM Tris-MES pH 7.5, 0.5 mM DTT). Protein concentration was measured by Bradford assay as described in section 2.4.4 and the samples stored at −80 °C until further use.

2.4.10. *In vitro* PM H^+-ATPase Activity Assay

PM H^+-ATPase activity was determined by resuspending 5 μg of microsomal proteins (section 2.4.9) in 10 μl resuspension buffer and 40 μl of reaction buffer containing 25 mM Tris-HCl pH 6.3, 50 mM KCl, 5 mM $MgCl_2$, 5 mM $CaCl_2$, 0.1 mM ouabain (Thermo Fisher Scientific), in presence of 1 mM NaN_3, 100 nM concanamycin A (Sigma-Aldrich), 0.02% (v/v) Brij58 (Sigma-Aldrich), 0.5 μg/μl BSA and 10 mM DTT. The reaction was initiated by the addition of 1 mM ATP, proceeded for 2 hrs at room temperature, and stopped with 150 μl of stopping reagent (8% (w/v) ascorbic acid, 2.1% (v/v) HCl and 0.7% (w/v) ammonium molybdate). After 5 min, 150 μl of 2% (w/v) tri-sodium citrate, 2% (v/v) acetic acid were added. After incubation for 15 min at room temperature the reduced phosphomolybdenum complex was quantified at 640 nm using

Tecan Spark® based on a standard curve of 0.25 – 2 mM K_2HPO_4. ATPase activity was measured with and without 400 µM orthovanadate (Na_3VO_4), and the difference between these two activities was attributed to the PM H^+-ATPase.

2.4.11. In Solution Trypsin Digestion of Microsomal Proteins

Microsomal proteins (section 2.4.10) were predigested for three hours with the endoproteinase Lys-C (0.5 µg/µl; Thermo Fisher Scientific) at 37°C in 6 M urea, 2 M thiourea, pH 8 (Tris-HCl). After 4-fold dilution with 10 mM Tris-HCl (pH 8), the samples were digested with 4 µl Sequencing Grade Modified trypsin (0.5 µg/µl; GE Life Sciences) overnight at 37 °C. The reaction was stopped by addition of TFA to reach pH≤ 3 before further processing.

2.4.12. Phosphopeptide Enrichment

Phosphopeptides were enriched over titanium dioxide (TiO_2, GL Sciences) as described by Larsen et al. (2005) with some modifications. TiO_2 beads were equilibrated with 100 µl of equilibration solution (1 M glycolic acid, 80% (v/v) acetonitrile and 5% (v/v) TFA). Trypsin digested peptides (section 2.4.11) were mixed with the same volume of the equilibration solution and incubated for 20 minutes with 2 mg TiO_2 beads with continuous shaking. After brief centrifugation, the supernatant was discarded, and the beads were washed 3 times in 80% (v/v) acetonitrile, 1% (v/v) TFA. Then the phosphopeptides were eluted with 240 µl of 1% (v/v) ammonium hydroxide. Eluates were immediately acidified by adding 50 µl 10% (v/v) formic acid to reach pH < 3. Enriched phosphopeptides were desalted over C18-Stage tips (section 2.4.13) prior to mass spectrometric analysis (section 2.4.14). Only phosphopeptide-enriched fractions were analyzed.

2.4.13. Peptide Desalting over C18-Stage Tips

Prior to LC-MS/MS analysis the peptides were desalted over C18-Sage tips as described by Rappsilber et al. (2007) as follows:

C18-Stage tips were made by introducing two discs of C18 (Millipore, Schwalbach) to the bottom of a 200 µl pipette tip. The Stage-tips were conditioned by loading 50 µl of elution solution (80% (v/v) acetonitrile and 0.5% (v/v) acetic acid). They were placed in a 2 ml Eppendorf tube and centrifuged for 30 sec at 6000 g. After discarding the liquid, the tips were equilibrated 2 times by loading 100 µl of 0.5% (v/v) acetic acid and centrifugation as before. The acidified peptide sample (pH 2 – 3) was loaded. After centrifugation, the tips were washed two times with 100 µl 0.5% (v/v) acetic acid. Peptides were eluted by adding 20 µl of elution solution and the eluate was collected in a fresh 1.5 ml Eppendorf tube. The samples were concentrated in Savant SpeedVac Concentrator (SVC-100H Vacuum centrifuge, Bachofer, Reutlingen) and stored dry at -20°C until further analysis.

2.4.14. Liquid Chromatography Tandem Mass Spectrometry (LC-MS/MS)

Desalted peptide mixtures were analyzed by LC-MS/MS using nanoflow Easy-nLC1000 (Thermo Scientific) as an HPLC-system and a Quadrupole-Orbitrap hybrid mass spectrometer (Q-Exactive Plus, Thermo Scientific) as a mass analyzer. Peptides were eluted from a 75 µm x 50 cm C18 analytical column (PepMan, Thermo Scientific) on a linear gradient running from 4 to 64% acetonitrile in 240 min and sprayed directly into the Q-Exactive mass spectrometer. Proteins were identified by MS/MS using information-dependent acquisition of fragmentation spectra of multiple charged peptides. Up to twelve data-dependent MS/MS spectra were acquired for each full-scan spectrum recorded at 70,000 full-width half-maximum resolution. Fragment spectra were acquired at a resolution of 35,000. Overall cycle time was approximately one second.

Protein identification and ion intensity quantitation was carried out by MaxQuant version 1.5.3.8 (Cox and Mann 2008). Spectra were matched against the Tomato proteome (*Solanum lycopersicum*, ITAG2.4, 34725 entries) using Andromeda (Cox et al. 2011). Thereby, carbamidomethylation of cysteine was set as a fixed modification;

oxidation of methionine, N-terminal protein acetylation as well as phosphorylation of serine, threonine and tyrosine were set as variable modifications. Up to two missed cleavages (trypsin) were allowed: Mass tolerance for the database search was set to 12 ppm on full scans and 20 ppm for fragment ions. Multiplicity was set to 1. For label-free quantitation, retention time matching between runs was chosen within a time window of two minutes. Peptide false discovery rate (FDR) and protein FDR were set to 0.01, while site FDR was set to 0.05. The score threshold for phosphorylation site localization was set to 50. Hits to contaminants (e.g. keratins) and reverse hits identified by MaxQuant were excluded from further analysis. For increasing quantitative coverage in label-free quantitation "match between runs" was selected in MaxQuant with a tolerance of 0.75 minutes within 20-minute time windows throughout the run.

Raw files and identified spectra of the Systemin phosphoproteomics analysis were submitted to ProteomeXChange and are available to the public under the accession number PXD010819.

2.5. Mass Spectrometric Data Analysis and Statistics for the Systemin Phosphoproteomics Analysis

Reported ion intensity values were used for quantitative data analysis. cRacker (Zauber and Schulze 2012) was used for label-free data analysis of phosphopeptide ion intensities based on the MaxQuant output (evidence.txt). All phosphopeptides and proteotypic non-phosphopeptides were used for quantitation. Within each sample, ion intensities of each peptide ions species (each m/z) were normalized against the total ion intensities in that sample (peptide ion intensity/total sum of ion intensities). Subsequently, each peptide ion species (i.e. each m/z value) was scaled against the average normalized intensities of that ion across all treatments. For each peptide, values from the three biological replicates were averaged after normalization and scaling. In case of non-phosphopeptides, protein ion intensity sums were calculated from normalized scan scaled ion intensities of all proteotypic peptides.

2.6. *A. tumefaciens*-mediated Transformation of Tomato Plants

S. lycopersicum (cv. UC82B) was transformed with *A. tumefaciens* harboring the RLKs CRISPR/Cas9 constructs (section 2.1.10.1) as described by Bosch et al. (2014b).

Tomato seeds were exposed to a heat treatment of 70°C overnight. The following day, the seeds were washed for 5 min in 70% (v/v) EtOH and incubated for 3 hrs in 10% (w/v) trisodium phosphate to inactivate any viruses contained in the seeds and thus prevent infection of the plant material. After washing the seeds three times with ddH_2O for 5 min, they were sterilized for 10 min in 1.5% (v/v) sodium hypochlorite solution. Another 5 washes were performed with about 100 ml sterile ddH_2O, before seeds were sown on germination medium (section 2.1.6.3) and incubated for 11 days in the dark at 22°C.

The cotyledons of etiolated tomato seedlings were cut and placed upside down on conditioning medium (section 2.1.6.3) in the dark at 26°C for two days. Subsequently, they were co-cultivated with *A. tumefaciens* (OD_{600} 1.0) harboring the corresponding construct resuspended in co-cultivation medium (section 2.1.6.3). After co-cultivation for two days in the dark at 26°C, the cotyledons were transferred to kanamycin-containing selection medium (section 2.1.6.3) for selecting transformed explants. The explants were grown for 1 week on 35 mg/l Kanamycin selection medium followed for 2 weeks on 50 mg/l Kanamycin selection medium. Then, until generation of shoots on 100 mg/l Kanamycin selection medium (5 to 6 weeks). When green shoots appeared, they were separated from the remaining browning cotyledons and placed in direct contact with the selection medium. When the shoots were about 2-3 cm tall, they were transferred to rooting medium (section 2.1.6.3). When the roots were well developed, the regenerated plantlets were tested for the presence of the CRISPR/Cas9 construct by PCR using the primers in Table 2.10. In addition, they were tested for the occurrence of mutations in the corresponding RLK by native PAGE (section 2.3.4).

Both transgenic and mutated transgenic plants were transferred to soil and grown for seed collection in the greenhouse at a 16-h photoperiod with supplemental light and a 26°C/18°C day/night temperature regime. Plants were fertilized at weekly intervals (GABI plus 12-8-11; N, P, K fertilizer at 2 ml/l).

T_0 transgenic plants were obtained for all RLKs CRISPR/Cas9 constructs except for PORK1, whose transformation was still running at the time of submission of this thesis. The number of T_0 plants for the different RLK CRISPR/Cas9 constructs was: 12 plants for SYR2, 5 plants for LRRXIV, 4 plants for GHR1, 3 plants for LYK4 and 1 plant for each of PSKR2, LRK10L1.2 and SYR1.

Mutations were further analyzed in the T_1 generation. To this end, virus-inactivated T_1 seeds of the transgenic tomato plants were grown in soil, for 14-16 days at a photoperiod of 16 h, 26°C and 100 μmol m^{-2} s^{-1} light intensity in a growth chamber (Conviron Adaptis C, Manitoba, Canada). These plants were genotyped for mutations in the corresponding RLKs by PCR amplification of genomic fragments spanning the targeted mutation sites using the primers listed in Table 2.11. Restriction digests (section 2.3.6) and/or native PAGE (section 2.3.4) were performed to detect any mutations. PCR products that showed homozygous mutation patterns were sent to sequencing (section 2.3.16) to confirm the type of mutation.

Confirmed homozygous T_1 mutants were tested for the presence of the CRISPR/Cas9 constructs by PCR using the primers listed in Table 2.10. Only construct-free (Cas9-negative) homozygous mutants were grown further in the green house for seed collection.

2.7. Detection of the Expression Level of Candidate RLKs after Wounding

Three homozygous lines of tomato plants overexpressing Prosystemin and double-mutated Prosystemin (section 2.2.3) as well as wild type were used for this experiment. After heat treatment and virus inactivation as described in section 2.6, the seeds were sown in soil and grown for 14 days in a growth chamber (Conviron Adaptis) at 26°C and at a photoperiod of 16 hrs, 26°C and 100 μmol m^{-2} s^{-1} light intensity. For the zero hr time point, samples were collected from 6 plants per line, pooled in a reaction tube, transferred to liquid nitrogen and stored at -80°C. The plants were then wounded twice on a leaflet of the first leaf with a hemostat. After 8 hrs, the wounded (local) leaves of 6 plants from each line were collected, combined in a reaction tube, transferred to liquid nitrogen and stored at -80°C. In the following, the RNA was isolated (section 2.3.17) and cDNA was synthesized (section 2.3.18). The expression level of the candidate RLKs was measured by qRT-PCR (section 2.3.19) using the primers listed in Table 2.6.

2.8. Detection of *PI-II* Expression in RLK Knocked-out Tomato Plants after Wounding

For this experiment, two T_2 lines homozygous for the mutation and transgene-free plants were used for SYR2, LRRXIV, LYK4 and GHR1. Only one such line was used for PSKR2, LRK10L1.2 and SYR1. The plants were grown, and the wounding experiment was performed as described in section 2.7. For this experiment unwounded (systemic) leaves were collected 8 hrs after wounding in addition to the wounded (local) leaves. Afterwards, the RNA was isolated (section 2.3.17) and cDNA was synthesized (section 2.3.18). The expression level of *PI-II* was measured by qRT-PCR (section 2.3.19) using the primers listed in Table 2.6.

2.9. Transformation of *S. peruvianum* Cells by Particle Bombardment

The CRISPR/Cas9 constructs designed to knock out the candidate RLKs were transformed into *S. peruvianum* cells by biolistic transformation as described by Cedzich et al. (2009).

Six-day old *S. peruvianum* cells were transferred from the cell suspension culture as a thin layer on sterile filter paper placed on 10% (w/v) sucrose-containing Nover solid medium (section 2.1.6.2). They were incubated in the dark for 3 hrs at 26°C. During this time, 60 mg of gold particles (1 μm; BioRad) were washed for 10 min in 70% (v/v) EtOH, then 3 times with sterile ddH$_2$O and resuspended in 1 ml of 50% (v/v) sterile glycerol solution. The purified recombinant binary vectors harboring the RLK targeting CRISPR/Cas9 constructs were linearized by restriction digestion with *PmeI* (Thermo Fisher Scientific), which does not cut the binary vector in the Cas9 or Kanamycin coding sequences nor in the gRNAs. The linearized vectors were purified and adsorbed on the prepared gold particles as described by Klein et al. 1988 as follows: 50 μl gold particles (about 3 mg) were washed 2 times with 200 μl of sterile ddH$_2$O. After brief centrifugation and discarding the solutions, the gold particles were resuspended in 10 μl of the linearized purified binary vector and mixed by vortex for 1 min then placed on ice. To the side of the tubes 20 μl of 0.1 M spermidine and 50 μl of 2.5 M CaCl$_2$ were added in a way that they do not mix with each other before coming in contact with the gold particles. The solutions were mixed by vortex for 3 min. After centrifugation the loaded Gold particles were washed two times with 250µl of 70% (v/v) EtOH and were resuspended in 120 µl of 99.9% (v/v) EtOH.

To the preincubated *S. peruvianum* cells, 24 µl of DNA-coated gold particles per plate were delivered using a PDS-1000/He™ Biolistic Particle Delivery System (BioRad) under 1100 psi partial vacuum and a distance of 9 cm. After bombardment, the filter papers containing the cells were transferred to antibiotic-free solid Nover medium (section 2.1.6.2) for two days recovery in the dark at 26°C. The cells were then transferred to solid selection Nover medium (section 2.1.6.2) supplemented with 75 mg/l Kanamycin. Transformed calli started to grow about 3 to 4 weeks after transformation. They were subcultured separately every two weeks on the same selection medium. Liquid cultures were established from the calli that were confirmed by PCR to carry the CRISPR/Cas9 constructs, and to have mutations in the target sequence of the corresponding RLK as confirmed by restriction digest (section 2.3.6) and/or native PAGE (section 2.3.4). The liquid cultures were subcultured several times until the optimal cell density was reached.

For genotyping of the knocked-out cell cultures the PCR products spanning mutation sites were cloned into pCR 2.1-TOPO® vector (Thermo Fisher Scientific). About 8 clones per cell culture were sent to sequencing for mutations detection.

2.10. Alkalization Response of *S. peruvianum* Cell Suspension Cultures

Ten ml of 7-day old *S. peruvianum* cell suspension cultures were transferred to scintillation vials on an orbital shaker. pH was measured continuously using INLAB Semi Micro electrodes (Mettler Toledo, Gießen) with the Seven-Multi pH Meter (Mettler-Toledo) and LabX direct pH 2.1 software (Mettler Toledo). The cells in the scintillation vials were equilibrated for at least 1 hr before starting each experiment. Depending on the experiment, Systemin, A17 or Flg22 peptides (Table 2.4) were added in aqueous solution of a total volume of 50 µl and in a final concentration of 10 nM. The pH change in the medium was then recorded over a period of 1 hr every 10 sec and the data were analyzed by Microsoft Office 365 Excel.

2.11. Detection of Stomatal Movement in *ghr1* T$_2$ lines in Response to Wounding

Changes in stomatal aperture in response to wounding were analyzed in 2 *ghr1* T$_2$ lines and compared to wild type. Three 3-weeks old plants per line were wounded twice using a hemostat. After 30 minutes, the wounded leaves were collected. Unwounded leaves were collected from the same plants prior to wounding as controls. Stomatal impressions were obtained by applying a thin layer of colorless nail polish (Kiko, France) to the lower side of the collected leaves. After drying the impressions were transferred to a microscope slide using transparent sticky tape (tesa, Norderstedt) and analyzed with a Axio Imager Z.1 stereomicroscope, equipped with the AxioVision RS software and the AxioCam MRm (Zeiss, Jena, Germany). Five photos of different leaf areas were taken at 400 x magnification with differential interference contrast (DIC). The photos were analyzed using ImageJ software v. 1.52a (Schneider et al. 2012). For each line the stomatal aperture index (width/length) and stomatal size (width*length) were measured for 50 different stomata. Stomatal density was calculated as the number of stomata per mm^2 per line.

2.12. Statistics

Microsoft Office 365 Excel was used for the statistical analysis. The statistical tests and significance levels used are given in the figure legends.

3. Results

3.1. Phosphoproteomics Analysis of the Microsomal Membrane Protein Fraction of Systemin-treated *S. peruvianum* Cell Suspensions

3.1.1. Choice of Controls

An experiment was performed to test the early cellular responses of *S. peruvianum* cell suspensions to Systemin and compare it with that of the inactive Systemin analog A17, which has an Ala substitution at position 17 (Thr) in the peptide sequence (Pearce *et al.*, 1993). The ability of A17 to competitively antagonize the alkalization response of cell cultures to 10 nM Systemin was tested by adding increasing concentrations of A17 (1 nM, 10 nM, 100 nM and 1000 nM), and the pH change was monitored continuously over time.

Stimulation of the cell culture with 10 nM Systemin resulted in a rapid alkalization of the medium from pH 5.5 to pH 6.4 within about 8 minutes (Figure 3.1 A). Treatment with A17 alone could not trigger this alkalization response. Addition of increasing concentrations of A17 to the Systemin treatment significantly reduced the alkalization response, thus confirming A17 as a potent competitor of the Systemin response (Figure 3.1B). The inactive A17 Systemin analog was therefore used as a control in the proteomics experiment, to later filter the data set for Systemin-specific responses.

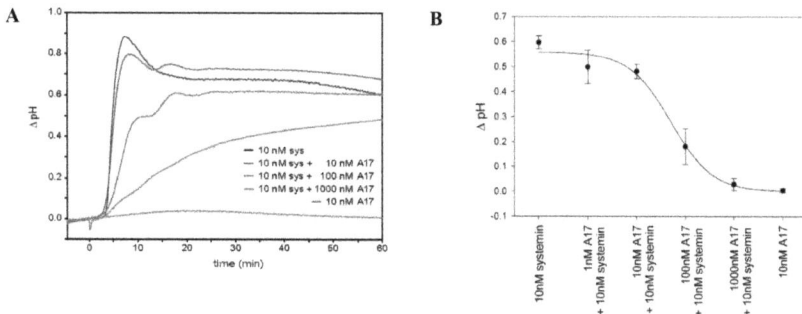

Figure 3.1: Competitive Effect of A17 on Native Systemin Response in *S. peruvianum* Cell Suspension Cultures.

(**A**) The alkalization response of *S. peruvianum* cell cultures measured continuously over time. A seven-day-old cell culture was treated separately with 10 nM Systemin (blue), 10 nM A17 (red) and 10 nM Systemin with increasing concentrations of A17 (10 nM, 100 nM and 1000 nM) (purple to pink) and pH change was monitored over time. (**B**) Inhibition curve of A17. ΔpH the difference in pH at 0 min (just before adding the peptide) and 8 min (when the maximum was reached) is shown for different peptide treatments. Data represent the mean of three independent experiments using three independent batches of cell cultures. Error bars represent the standard deviation.

3.1.2. Experimental Design

Based on the results of the above experiment, A17 was used as negative control for Systemin treatment in the phosphoproteomics experiment. Water was used as a mock control. Thereby, the experiment was performed as follows:

Seven-day-old cell suspension cultures (200 ml) of *S. peruvianum* were subjected to stimulation with 10 nM Systemin (Sys), 10 nM A17 (A17), or treated with equal volume (1 ml) of water (W). For each treatment, a batch of cells was harvested after 2, 5, 15 and 45 minutes over a sieve using vacuum and were frozen immediately in liquid nitrogen (Figure 3.2). Untreated cells were harvested as controls. All treatments and time points were sampled from three independent biological replicates using three independent sets of cells for each experiment. For phosphopeptide profiling, microsomal membranes were isolated from each sample, digested with trypsin, enriched for phosphopeptides and subjected to untargeted data-dependent acquisition by LC-MS/MS. Data analysis was carried out jointly for all samples by averaging quantitative information for each time point and treatment across the three replicates.

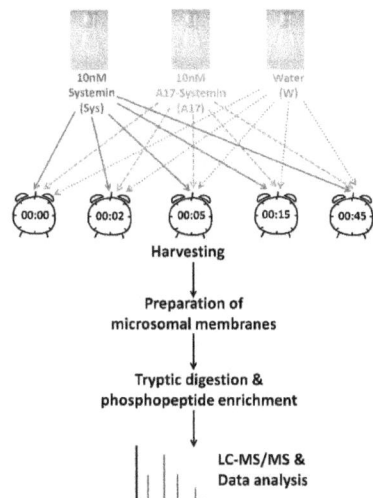

Figure 3.2: Experimental Design and Analytical Workflow of the Phosphoproteomics Experiment.
Seven-day-old *S. peruvianum* cell cultures were treated with 10 nM Systemin, 10 nM A17, and water separately. Cells were harvested after 2, 5, 15 and 45 mins after treatment. Untreated cells were harvested as 0 min control. The experiment was repeated three times using three independent batches of the cell culture. The microsomal membrane fraction was isolated by differential centrifugation, proteins were trypsin-digested and the phosphopeptides were enriched. These were subjected to untargeted data-dependent acquisition by LC-MS/MS using nanoflow Easy-nLC1000 (Thermo Scientific) as an HPLC-system and a Quadrupole-Orbitrap hybrid mass spectrometer (Q-Exactive Plus, Thermo Scientific) as mass analyzer.

3.1.3. Description of the Phosphoproteomics Data Set

After quantitative data analysis by cRacker (Zauber and Schulze 2012), a total of 3312 phosphopeptide species were identified (see Appendix A.1 for a complete list of identified phosphopeptides). Most of the peptides identified were singly phosphorylated (about 73%). About 13% were phosphorylated at two sites and very few were phosphorylated at three different positions. From the different phosphorylation events 89.5% were phosphoserines, about 10% were phosphothreonines, and 0.4% were phosphotyrosine (Figure 3.3 A and B), which is comparable with percentages of phosphoresidues identified in previous studies (Ghelis, 2011; van Wijk et al. 2014). Quantitative information for at least one treatment was obtained for 2960 phosphopeptide species matching 1309 protein models in at least one biological replicate. Also, for each of the three treatments the majority of phosphopeptides were identified at all time points (Figure 3.4 A). This high overlap indicates high analytical reproducibility and high potential for good quantitative coverage. In addition, at each sampled time point, the majority of phosphopeptides was identified under all treatment conditions (Figure 3.4 B) except at 5, 15 and 45 minutes, where a large number of phosphopeptides was identified only after Systemin stimulation. By far fewer specifically phosphorylated peptides were identified under A17 and water control treatments.

A
B

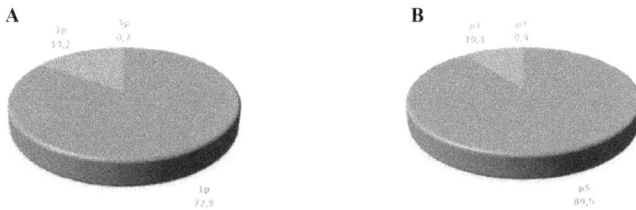

Figure 3.3: Frequency Distributions of the Identified Phosphopeptides.
(A) The Frequency distribution of the phosphorylated residues in the 3312 identified phosphopeptides. The majority of these peptides were phosphorylated at one position (about 73%). 13% of them were phosphorylated at two positions and only very few at three positions. (B) Distribution of the phosphorylated residues in the identified phosphopeptides. From the 3836 identified phosphorylation sites 89.5% were phosphoserines, 10.1% were phosphothreonines and 0.4% were phosphotyrosines. These pie charts were made depending on the phosphopeptides as determined by MaxQuant, which are listed in appendix A.1.

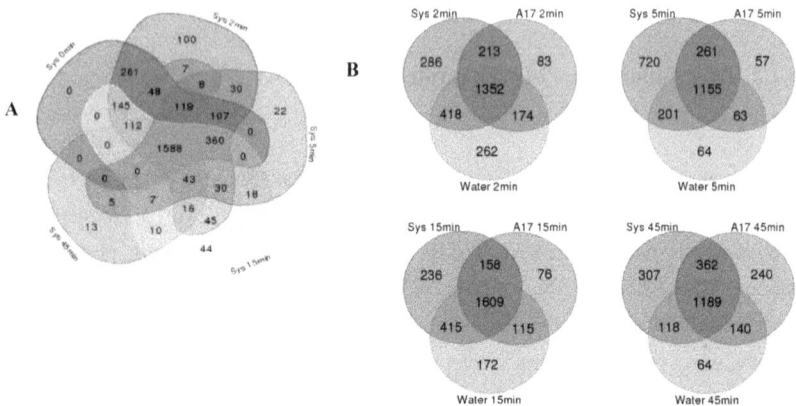

Figure 3.4: Overview of Identified Phosphorylation Sites with Respect to Different Time Points and Treatments.

Venn diagrams were drawn using the http://bioinformatics.psb.ugent.be/webtools/Venn/ website for the phosphopeptides listed in appendix A.1. All phosphopeptides were included that had a normalized intensity at specific time point and specific treatment. (**A**) All identified phosphopeptides at different time points under one treatment condition (Systemin). Most of the phosphopeptides (1588 peptides) were identified under all time points. (**B**) Overlap of phosphopeptides identified at different time points upon Systemin, A17 and water treatments. After 5, 15 and 45 min, larger number of phosphopeptides were identified after Systemin treatment in comparison to A17 and water treatments.

3.1.4. Phosphorylation Time Profiles

Treatment of cell suspension cultures with Systemin resulted in characteristic changes of phosphorylation patterns. k-means clustering was used to group the phosphorylation profiles of all detected phosphopeptides upon different treatments over time (Figure 3.5; Appendix A.2).

As shown in Figure 3.5 A, once Systemin was applied to cells a very rapid dephosphorylation of 827 phosphopeptides occurred (cluster A), another 465 phosphopeptides showed a transient increase in phosphorylation after 2 minutes (cluster B), 501 phosphopeptides showed a transient increase in phosphorylation after 5 minutes (cluster C), and 705 phosphopeptides displayed a transient peak of phosphorylation at 15 minutes (cluster D). Further 1001 phosphopeptides were classified as not responsive to Systemin (cluster F).

Similar patterns of k-means clustering resulted after A17 stimulation (Figure 3.5 B). However, under A17 treatment no phosphopeptides displayed a transient increase in phosphorylation at 5 minutes. Instead, 1042 phosphopeptides showed an increase in phosphorylation at 45 minutes (cluster E). Such response was not observed upon Systemin treatment.

Stimulation of the suspension cells with water resulted in response groups very similar to clusters observed upon A17 treatment (Figure 3.5 C). Particularly, also cluster C with a phosphorylation peak at 5 minutes was missing and cluster E with maximum phosphorylation at 45 minutes was observed.

The missing cluster C and emergence of a "later" maximum in cluster E in both A17 and water treatments suggests that the control treatments A17 and water do trigger some cellular response, but weaker and delayed compared to Systemin treatment.

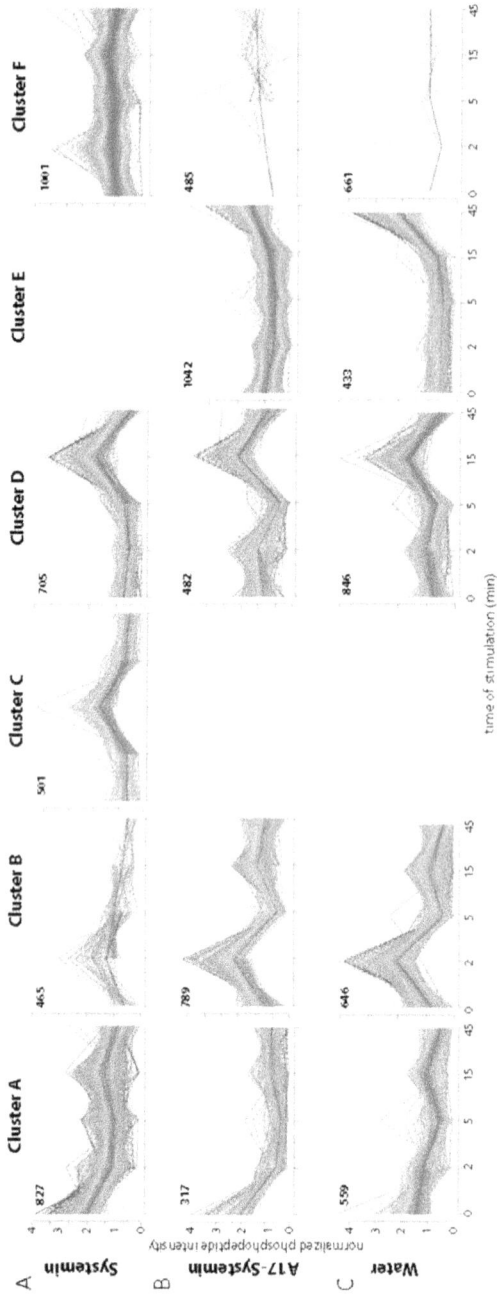

Figure 3.5: *k*-means Clustering of Phosphorylation Profiles.

Phosphorylation time profiles of identified phosphopeptides upon (**A**) Systemin, (**B**) A17 and (**C**) Water treatment subjected to *k*-means clustering using cRacker. Numbers of identified phosphopeptides are indicated within each panel.

3.1.5. Verification of the Phosphorylation Time Profiles Upon Different Treatments

After Systemin perception, the alkalization of the extracellular milieu, which leads to alkalization of the culture medium is caused by the inactivation of the PM H^+-ATPase (Schaller and Oecking 1999). Therefore, this proton pump was used as a biological model to verify the fidelity of the observed phosphorylation time profiles under different treatments. A closer look was thus taken at the phosphorylation profiles of peptides derived from the PM H^+-ATPase in the data set (Figure 3.6 A and B).

In addition, in order to validate the phosphorylation profiles of the PM H^+-ATPase, its activity was measured for the different time points under different treatments (Figure 3.6 C).

For the PM H^+-ATPase activity assay, microsomal proteins were extracted from the samples collected after the different time points for each treatment and the buffer for resuspension was chosen to maintain their native conformation and activity. Five μg of extracted proteins were diluted in the reaction buffer containing 1mM ATP and used for the activity assay. A colorimetric assay was used to detect the free inorganic phosphate released by ATP hydrolysis. The reaction was performed twice; once with vanadate (a specific inhibitor of P-type ATPases) and once without it. The difference between both measurements was attributed to the PM H^+-ATPase activity.

It is well known that PM H^+-ATPase plays a critical role in metabolite uptake, response to the environment, growth, and development through maintaining cellular membrane potential in plants (Falhof et al. 2016; Haruta et al. 2015). Its activity is controlled by (de)phosphorylation of several Ser and Thr residues within the carboxy terminal regulatory domain. One major regulatory site is the penultimate Thr residue, phosphorylation of which results in proton pump activation (Elmore and Coaker 2011; Haruta et al. 2015).

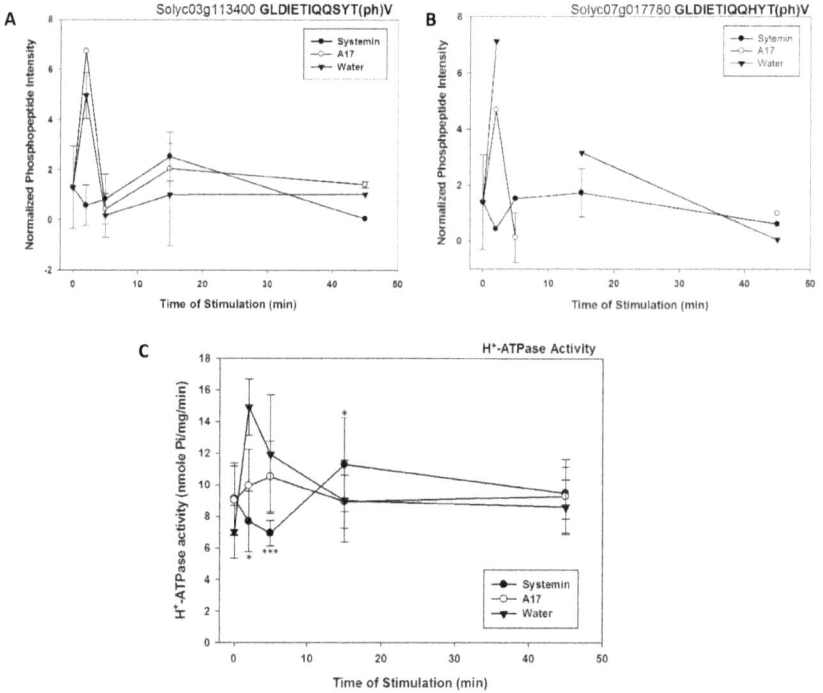

Figure 3.6: PM H⁺-ATPase Phosphorylation Profiles and Activity over Time after the Different Treatments.
Phosphopeptides harboring the penultimate C-terminal activation site of PM H⁺-ATPases Solyc03g113400 (**A**) and Solyc07g017780 (**B**) were identified with different time profiles under all three treatments. Systemin treatment induced early dephosphorylation of this site in both peptides, while A17 and water induced its phosphorylation. Mean values of 1-3 biological replicates are shown with standard deviation. (**C**) H⁺-ATPase activity assay upon treatment with Systemin, A17, and water. The values represent the mean (n=12) of three independent biological replicates and four technical repeats for each measurement. Error bars represent standard deviation. Differences between Systemin, A17 and water treatments were tested for significance by two-tailed t-test. * significant difference at $p<0.05$. *** significant difference at $p<0,001$. Black dots: Systemin treatment; white dots: A17 treatment; black triangles: water treatment.

Two phosphopeptides containing the penultimate C-terminal activation site of two PM H⁺-ATPases (Solyc03g113400 and Solyc07g017780) were identified with different time profiles under all three treatments. In agreement with the observed alkalization of the growth medium under Systemin treatment the PM H⁺-ATPase was slightly dephosphorylated at 2 minutes after treatment (Figure 3.6 A and B). This was associated indeed with a measurable decrease in H^+-ATPase activity at 2 to 5 minutes of Systemin treatment (Figure 3.6 C). The phosphorylation status of this activation site increased after 15 minutes of Systemin treatment, which was accompanied by an increase in the H^+-ATPase activity at that time point. The data indicate that Systemin perception causes temporal inactivation of two isoforms of the PM H^+-ATPase by dephosphorylation of their penultimate C-terminal activation site. In contrast, A17 and water treatments induced a transient increase in phosphorylation of these sites and H^+-ATPase activity at 2 minutes, which explains the increase of H^+-ATPase activity at 2 and 5 minutes (Figure 3.6 C). These observations are in agreement with the k-means clustering of these phosphopeptides. They were clustered in cluster D (phosphorylation maximum at 15 minutes) after Systemin treatment. For A17 and water treatments, they were found in cluster B (phosphorylation maximum at 2 minutes). The data support the conclusion that the phosphorylation time profiles presented in Figure 3.5 reflect precise quantitative changes in phosphorylation status causing actual measurable effects on protein activities.

3.1.6. Processes Involved in the Systemin Response

To find out which biological processes are influenced by Systemin treatment, an over-representation analysis (Fisher Exact Test) was performed for phosphopeptides present in the Systemin clusters A to F (Appendix A.3; Figure 3.7).

This analysis revealed that especially proteins with functions in signaling (Mapman bin 30; $p < 3.7E-19$), transport (Mapman bin 34; $p < 2.3E-24$), cellular processes (Mapman bin 31; $p < 3.0E-16$) and the cell wall proteins (Mapman bin 10; $p < 3.9E-5$) were highly over-represented in all clusters. Furthermore, several phosphopeptides potentially involved in stress responses (Mapman bin 20), protein-related responses (Mapman bin 29) and RNA-related processes (Mapman bin 27) were found to be over-represented in cluster A (rapid dephosphorylation) and cluster D (phosphorylation maximum at 15 minutes). Since Systemin signaling induces the production of ET and JA (Degenhardt et al. 2010; Felix and Boller 1995) the phosphopeptides assigned to hormone signaling proteins (Mapman bin 17) also were considered for further analysis, although they were not significantly over-represented in any of the Systemin response clusters.

It becomes apparent from Figure 3.7 that the abundance distribution of proteins in different functional sub-bins revealed cluster-specific distributions. For example, cellulose synthesis proteins (bin 10.2) and proteins involved in auxin signaling (bin 17.2.3) were most abundant in cluster B (2 min). Biotic stress-response proteins (bin 20.1), calcium signaling proteins (bin 30.3) and protein degradation (bin 29.5) were most abundant in cluster C (5 min). The P-/V-ATPases (bin 34.1) and G-protein signaling (bin 30.5) were most abundant in cluster D (15 min). Vesicle transport proteins (bin 31.4) were highly abundant in clusters B and C, while transcription factors (bin 27.3) and general protein modification processes (bin 29.4) were highly represented in all clusters.

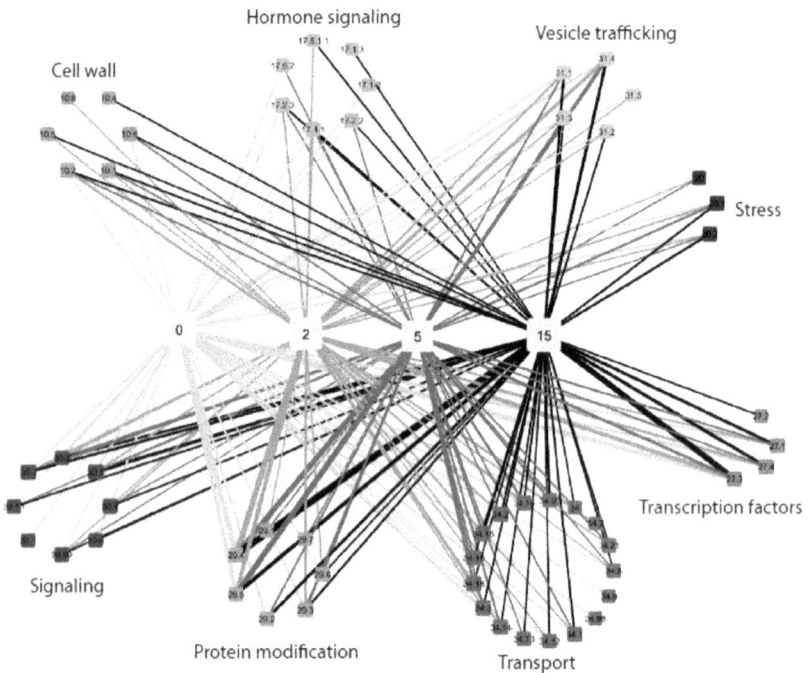

Figure 3.7: Over-representation Analysis of Functional Groups in each of the Systemin Clusters. A network view of processes over-represented at different time points after Systemin treatment. Yellow colored "time"-nodes represent k-means clusters (Figure 3.5) with maximum phosphorylation at respective time points. The "bin"-nodes represent the processes over-represented in the respective k-means clusters. The width of the edges represents the percentage of proteins of each bin within each k-means cluster. Edge color indicates the processes involved at different time points. The network was constructed using Cytoscape v. 3.7.1 software https://cytoscape.org/.

3.1.7. Filtering Out Systemin-specific Responses

To define precisely which phosphopeptides of the proteins in the Systemin-stimulated processes showed a Systemin-specific response, the phosphorylation time profiles for each phosphopeptide induced by Systemin were compared with those obtained upon treatment with A17 (Figure 3.8). Since the mock treatment with water resulted in responses very similar to A17 treatment, the water treatment was not further considered for definition of Systemin-specific responses.

The phosphopeptides identified after A17 treatment were divided into 4 groups: 'equal', 'earlier', 'later' and 'not in A17'. Those phosphopeptides that exhibited a phosphorylation peak at the same time point after A17 and Systemin treatment were classified 'equal'. If the phosphopeptides peaked earlier after A17 as compared to Systemin treatment, they were grouped in the 'earlier' group. The phosphopeptides found in an A17-cluster with a phosphorylation peak at a time point later than Systemin treatment were grouped in the 'later' group. Those phosphopeptides that were not identified after A17 treatment at all were considered 'Not in A17'.

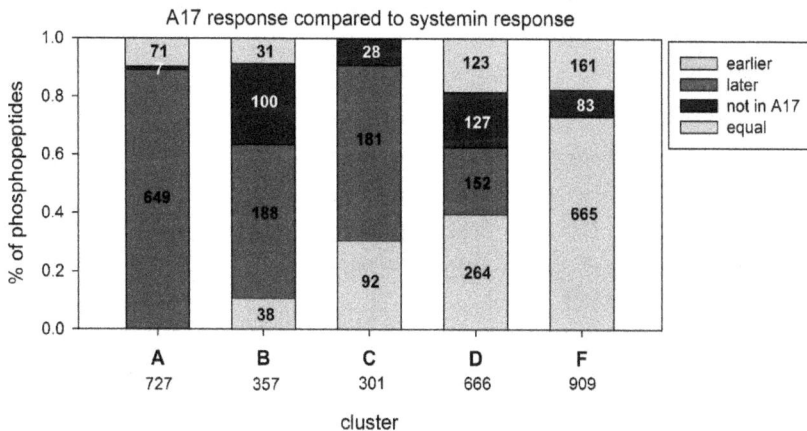

Figure 3.8: Comparison of Systemin and A17 Response.
The amount of phosphopeptides with a time-shifted response in A17 treatment compared to Systemin treatment is displayed. This bar graph includes only the phosphopeptides of Systemin clusters that belong to the proteins in the Systemin-stimulated processes (Figure 3.7). The phosphorylation time profiles of these phosphopeptides were compared after Systemin and A17 treatments. "Late" peptides display a later phosphorylation peak upon A17 stimulation. "Not in A17" peptides were not identified after A17 treatment. "Early" peptides upon A17 have a phosphorylation peak at earlier time points

compared to Systemin stimulation. "Equal" are peptides that showed a phosphorylation maximum at same time point after Systemin and A17 stimulation. Numbers indicate absolute numbers of classified phosphopeptides.

Out of 727 phosphopeptides in Systemin-induced cluster A, 649 were classified in a "later" cluster (B to F) upon A17 treatment, and seven phosphopeptides were not identified at all upon A17 treatment (Figure 3.8). Likewise, of the 357 phosphopeptides in Systemin-specific cluster B, 188 and 100 phosphopeptides showed a later phosphorylation peak or were not identified after A17 treatment, respectively (Figure 3.8). Since cluster C (phosphorylation maximum at 5 min) is only restricted to Systemin stimulation, most of the phosphopeptides were Systemin-specific; 181 phosphopeptides showed later phosphorylation peaks and 28 were not identified after A17 treatment. In cluster D, 152 were identified later and 127 phosphopeptides were not identified upon A17 stimulation. In contrast, 665 phosphopeptides that clustered as "non-responsive" upon Systemin treatment (cluster F) showed a phosphorylation peak in an earlier cluster in A17 treatment, 161 phosphopeptides were found with equal classification in A17 and Systemin, and 83 phosphopeptides were not identified in the A17 treatment. In total, 1515 phosphopeptides, which is the sum of all phosphopeptides that showed later phosphorylation peak or were not identified after A17 treatment in all Systemin-specific clusters, were classified as Systemin-specific. This makes about 45% of all identified phosphopeptides.

Based on this phosphopeptide classification, an over-representation analysis (Fisher Exact Test) was performed to find out which functional processes are over-represented in the "Systemin-specific" phosphopeptides in the 'later' and 'not in A17' groups (Appendix A.4). In both groups, proteins belonging to signaling (Mapman bin 30), transport (Mapman bin 34), cellular processes (Mapman bin 31), DNA-related processes (Mapman bin 28), and the cell wall proteins (Mapman bin 10) were significantly over-represented. Within the group of phosphopeptides not identified upon A17 treatment, proteins of vesicle transport (bin 31.4), cellulose synthesis (bin 10.2), P- and V-ATPases (bin 34.1), and auxin-related processes (bin 17.2.3) were highly enriched (Appendix A.4).

3.1.8. Kinases and Phosphatases Specifically Induced by Systemin

Since kinases and phosphatases play crucial roles in signal transduction pathways (Ardito et al., 2017; Ho, 2015), the Systemin-specific phosphopeptides, which were later or not at all identified after A17 treatment, were searched for peptides belonging to kinases and phosphatases proteins. Among the Systemin-specific phosphopeptides, 105 and 18 phosphopeptides were found matching 78 protein kinases and 18 phosphatases respectively (Appendix A.5). The phosphorylation time profiles of 56 kinase and 17 phosphatase phosphopeptides were used to construct a Pearson correlation network of Systemin-specific responses (Appendix A.6). From the initially defined 78 Systemin-specific kinases, not all phosphopeptides met the requirement of being detected with quantitative values in at least four samples across the time course and were therefore excluded from further analysis. Each phosphopeptide of a kinase or phosphatase was then correlated against all other phosphopeptides from the functional categories over-represented in each of the response clusters induced by Systemin (Figure 3.7). A stringent cutoff-value of r=0.85 was used to display individual proteins connected by the Systemin-specific kinases or phosphatases (Figure 3.9 A and B).

Specifically, for the proteins in cluster A with fast dephosphorylation response, a MAP-triple kinase (S(ph)SNATVEASGLSR, Solyc11g033270), a protein phosphatase 1 regulatory subunit (LSS(ph)PAAQS(ph)PSVSTK, Solyc02g070260) and a protein phosphatase 2 C (LTNS(ph)PDVEIK, Solyc10g005640) were the phosphopeptides with highest degree (number of interactions) in the correlation network (210, 134, 118 respectively). These proteins had highest numbers of Systemin-specific correlation partners. Other kinases with high degree were a MAP-triple kinase (AIS(ph)LPS(ph)SPR, Solyc02g076780, degree 118), a calcium-dependent protein kinase (TES(ph)GIFR, Solyc10g078390, degree 115), and a receptor-like kinase (S(ph)FENEIR, Solyc09g072810, degree 110).

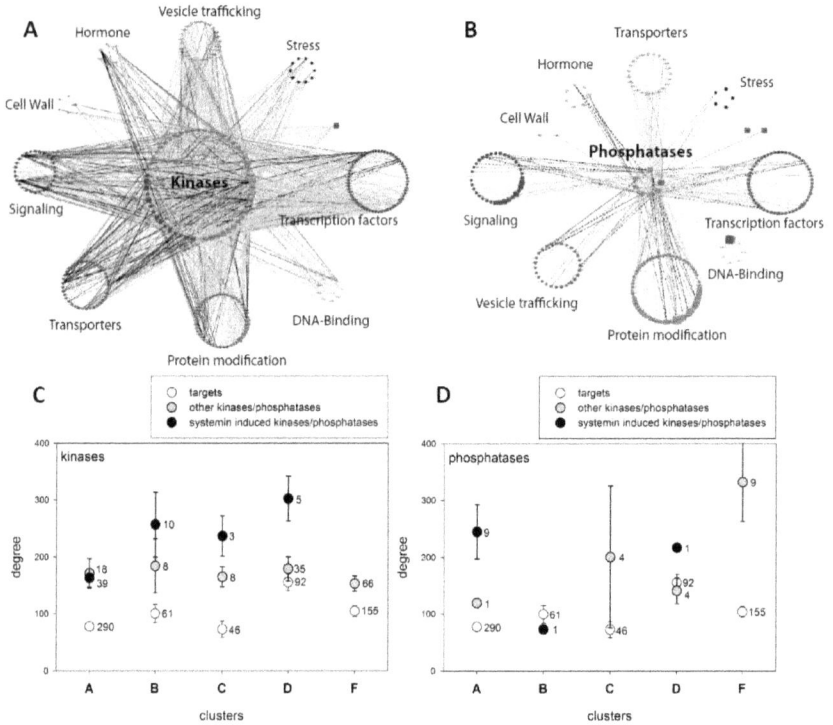

Figure 3.9: Correlation Network for Systemin-Specific Kinases and Phosphatases.
The phosphorylation time profiles of 56 kinase phosphopeptides and 17 phosphatase peptides were used to construct a Pearson correlation network with proteins of Systemin-specific responses. Each phosphopeptide of a kinase or phosphatase was correlated against all other phosphopeptides from the functional categories over-represented in each of the response clusters induced by Systemin (**A**) Kinase network. (**B**) Phosphatase network. Central nodes are the kinases or the phosphatases, peripheral nodes represent individual proteins summarized under Mapman sub-bins. Node size is proportional to the degree (number of interactions), edge width is proportional to the correlation r-value (cutoff r=0.85). (**C**) Degree distribution of Systemin-specific kinases, other kinases and their targets in different Systemin response clusters. (**D**) Degree distribution of Systemin-specific phosphatases, other phosphatases and their targets in different Systemin response clusters. Values represent means of degrees with standard error. Numbers of proteins in each class are indicated. The networks were constructed using Cytoscape v. 3.7.1 software https://cytoscape.org/.

Protein kinases with highest degree among proteins displaying transient phosphorylation at 2 minutes (cluster B) were CBL-interacting protein kinase

(T(ph)TCGTPNYVAPEVLSHK, Solyc04g076810, degree 112), Serine/threonine protein kinase (NQGS(ph)PSDTCSESDHK, Solyc01g108920, degree 110) and cell division protein kinase (DRLDS(ph)IDGK, Solyc07g063130, degree 102). For proteins with transient phosphorylation at 5 minutes (cluster C), the protein kinases with highest degree were the ET receptor (GGS(ph)QTDSSISTSHFGGK, Solyc05g055070, degree 103), and two receptor kinases (SIDLFTDVSEEGLS(ph)PR, Solyc09g083210, degree 49; ENPGDS(ph)GSLVQPGHDIEK, Solyc12g010740, degree 44). Kinases with highest degree among the proteins with transient phosphorylation at 15 minutes (cluster D) were the MAP triple kinase (SEEVDGS(ph)SSIR, Solyc11g033270, degree 210), the activating motif of mitogen-activated protein kinase 2 (MPK2) (VTSETDFM(ox)T(ph)EY(ph)VVTR, Solyc08g014420, degree 128), and CTR1-like protein kinase-3 (S(ph)AAGTPEWM(ox)APEVLR, Solyc09g009090, degree 122).

Interestingly, the protein kinases with high degree often were identified with several phosphorylation sites, which showed different response profiles during the Systemin treatment. For example, the MAP-triple kinase Solyc11g033270 (degree 210) was identified with two phosphorylation sites outside the kinase domain, S(ph)SNATVEASGLSR (Ser859) with rapid dephosphorylation (cluster A) and SEEVDGS(ph)SSIR (Ser345) with transient phosphorylation at 15 minutes (cluster D). The time profile of the dephosphorylated peptide S(ph)SNATVEASGLSR showed highest correlation with dephosphorylated peptides of an ubiquitin ligase (LEEGS(ph)SPEQR, Solyc05g007820), a YAK1 homologous protein kinase (TVYSY(ph)IQSR, Solyc03g097350) and a RNA recognition motif containing protein (SSDS(ph)QELTTTELK, Solyc02g062290). At 15 minutes of Systemin treatment, the other phosphopeptide of this protein kinase (Solyc11g033270) displayed a transient increase in normalized intensity. This correlated highest with transient phosphorylation of a calmodulin domain containing protein (DYS(ph)ASGYSSR, Solyc06g053530) and a transcriptional regulator (S(ph)PIPLEAVK, Solyc01g097890).

It was found that the dephosphorylated peptide S(ph)SNATVEASGLSR matching Solyc11g033270 in turn showed strong correlation with the peptide LTSS(ph)PEAEIK of a protein phosphatase 2C family protein Solyc10g005640 (r=0.97) suggesting kinase-phosphatase activation/inactivation loops upon Systemin stimulation.

Within this network (Figure 3.9 A and B), the degree distribution of phosphopeptides within Systemin-induced phosphorylation clusters (Figure 3.5) was displayed for Systemin-specific kinases/phosphatases, other identified kinases/phosphatases and the putative targets (Figure 3.9 C and D). Thereby, Systemin-specific kinases classified to clusters B, C, and D (which represented a transient phosphorylation increase), showed a higher degree than kinases without Systemin specificity, but kinases in general had higher degree than their target proteins. In contrast, for phosphatases, higher than average degree was observed for nine Systemin-specific phosphatases in cluster A (rapid dephosphorylation response). A higher than average degree was also observed for nine phosphatases in cluster F (no Systemin induced phosphorylation response). In conclusion, this suggests that Systemin-specific kinases exert their function particularly in transient phosphorylation responses observed in clusters with peaks at 2 minutes (cluster B), 5 minutes (cluster C) and 15 minutes (cluster D), while Systemin-specific phosphatases are particularly active triggering rapid dephosphorylation responses (cluster A).

3.2. Correlation of Systemin-Specific Kinases and Phosphatases with Major Elements of the Systemin Signaling Pathway

3.2.1. Plasma Membrane H+-ATPases

Phosphorylation profiles of phosphopeptides derived from two different PM H+-ATPases were highly correlated with those of phosphopeptides matching specific kinases and phosphatases (Figure 3.10). Phosphopeptides of a receptor kinase (Solyc02g079590; r=0.99), the MAP triple-kinase (Solyc11g033270; r=0.95) and MAP kinase 2 (MPK2; Solyc08g014420; r=0.98) were highly correlated to the phosphopeptide of the penultimate activating C-terminal phosphorylation site T955 (GLDIETIQQSYT(ph)V) of LHA1 (Solyc03g113400). On the other hand, LRR1 receptor-like kinase homolog (Solyc11g011020; r=0.94) was highly correlated to the phosphopeptide containing the activating phosphorylation site T964 (GLDIETIQQHYT(ph)V) of the AtAHA1 homolog Solyc07g017780 (Figure 3.10). This suggests that these kinases may directly phosphorylate the respective residues of the two H+-ATPases. The identification of MPK2 as a kinase involved in the systemin signaling pathway by correlation analysis is fully consistent with previous reports showing that MPK2, along with the closely related MPK1, are activated after mechanical wounding, herbivore attack, and Systemin application (Higgins et al. 2007; Holley et al. 2003; Kandoth et al. 2007). The C-terminal phosphopeptide of LHA1 also showed high correlation with POLTERGEIST-LIKE 5 (PLL5) phosphatase (Solyc06g076100), a member of the PP2C clade C phosphatases (Figure 3.10). Supporting the identification of PLL5 by correlation analysis, PP2C phosphatases have previously been implicated in the regulation of PM proton pump activity (Spartz et al. 2014), and in the regulation of PRR-induced signaling (Holton et al. 2015).

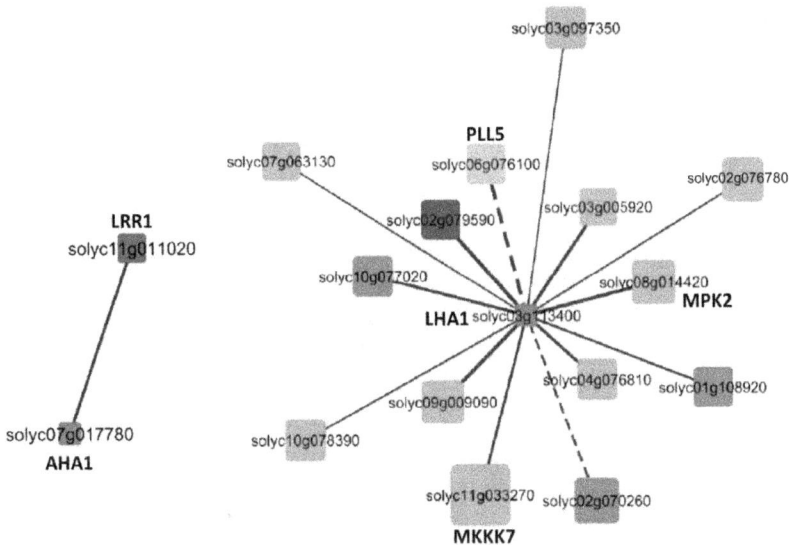

Figure 3.10: **Systemin-specific Kinase/Phosphatase Relationships with H⁺-ATPases LHA1 and AHA1 Homolog.**
A subnetwork is shown of the network for Systemin-specific kinases and phosphatases in Figure 3.9 A and B. Phosphopeptides of the H⁺-ATPases are shown along with their correlated Systemin-specific kinase/phosphatase partners. Solid edges represent kinase-substrate relationships and dashed edges represent phosphatase-substrate relationships. Node size indicates the degree in the overall network (Figure 3.9 A and B). Node color represents Mapman function. Purple: transport.p- and v-ATPases (34.1), blue: receptor kinase (30.2), cyan: phosphatases, pink: calcium-sensing kinases, orange: other kinases of protein.posttranslational modification category (29.4), gray: unclassified function. (35).

In order to confirm network predictions, the interaction of the C-terminal peptide of H⁺-ATPase LHA1 with candidate kinase MPK2 and candidate phosphatase PLL5 was tested *in vitro*.

3.2.1.1. Expression and Purification of PLL5 Phosphatase Domain and MPK2 Protein

The phosphatase domain of PLL5 (Solyc06g076100; amino acids 241-708) as well as the whole coding sequence of MPK2 (Solyc08g014420) were selected to be expressed in *E. coli*. Since MAP kinases must be activated before being used for *in vitro* assay, two forms of MPK2 were expressed; the wild-type form (WT) and a constitutively active mutant (CA) with two amino acid substitutions, D216G and E220A (Berriri et al. 2012; Genot et al. 2017).

The DNA sequences were amplified using wild-type tomato cDNA (extracted and synthesized from total leaf RNA) as template and cloned into pCR2.1-TOPO® (Thermo Fisher Scientific). After sequence confirmation (data not shown), the DNA fragments were digested using *NcoI* and *XhoI* and ligated into the expression vector pET21d (+) (Novagen) (Figure 3.11). The PLL5 phosphatase domain was cloned to be expressed as N-terminally His_6-tagged protein, while MPK2 (WT and CA) was cloned to be expressed as C-terminally His_6-tagged protein. The recombinant proteins were expressed in BL21 RIL *E. coli* (Stratagene) and then extracted and purified over Ni^{2+}-NTA agarose (Qiagen) via their His_6 tag (Figure 3.12).

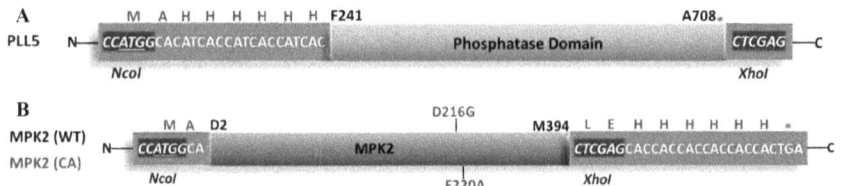

Figure 3.11: Scheme of Expression Constructs for the Phosphatase Domain of PLL5 and for MPK2.
(**A**) A scheme representing the expressed phosphatase domain (from amino acid 241 – 708) of PLL5. It was expressed as N-terminally His_6-tagged protein in *E. coli*. (**B**) scheme representing the expressed MPK2 protein as C-terminally His_6-tagged protein in *E. coli*. It was expressed in two forms the wild-type (WT) and the constitutively-active (CA) version, which has a double mutation of D216G/E220A shown in purple. Sequences that are not part of the original proteins are indicated in red. The start codon is underlined. Sequences marked in dark red are recognition sites of the restriction enzymes used for cloning. * represents the stop codon.

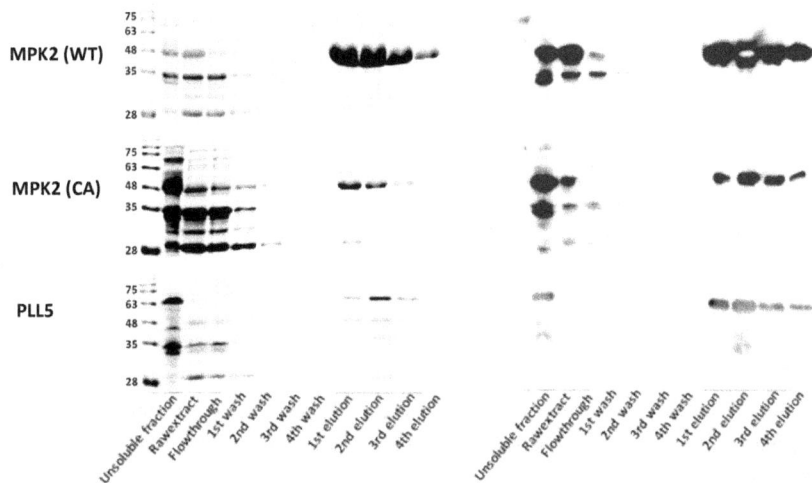

Figure 3.12: Purification of PLL5 Phosphatase Domain and MPK2 (WT and CA) using Ni^{2+}-NTA Agarose Column Chromatography.

The His-tagged PLL5 phosphatase domain and the WT and CA versions of MPK2 were expressed in *E. coli* (BL21 RIL) and purified using Ni^{2+}-NTA agarose. The left panels show different fractions of the purification procedure separated by 15% SDS-PAGE and stained with Coomassie Brilliant Blue. Corresponding western blots of the purified His-tagged proteins are shown in the right panels. Immunolabelling was performed using Anti-His tag antibodies and secondary rabbit-anti-mouse IgG peroxidase conjugate followed by enhanced chemiluminescence (ECL) detection.

All three proteins were purified successfully, with the recombinant protein as the major band after Ni^{2+}-NTA chromatography (Figure 12). The expected mass of the expressed PLL5 phosphatase domain is 54.4 kDa. The apparent mass of the purified PLL5 phosphatase domain was larger than expected; about 63kDa (Figure 3.12). Depending on their acidic amino acid content (Asp and Glu), proteins may run slower during SDS-PAGE and appear larger than expected (Guan et al. 2015; Whitfield et al. 1995). The difference between the predicted and apparent molecular weight shows a linear correlation with the percentage of acidic amino acids in a protein (Guan et al., 2015). For PLL5 phosphatase domain, the percentage of acidic amino acids is 17.4. Therefore, according to the equation of Guan et al. (2015), it's apparent mass will be about 8 kDa larger than expected (62.4 kDa), which corresponds well with the band observed for PLL5 in Figure 3.12.

For MPK2, the WT and CA forms, the expected calculated size is about 46.4 kDa. (Figure 3.12). Its acidic amino acids content is 13,4%, which according to Guan et al. (2015), would result in an increased molecular weight of 48.7 kDa, which is close to the calculated mass, and the apparent mass displayed in Figure 3.12.

3.2.1.2. PLL5 Phosphatase and MPK2 Kinase *in vitro* Assays

To establish the phosphatase assay, the purified His-tagged PLL5 phosphatase domain was incubated with the candidate phosphorylated substrate, i.e. the synthetic C-terminal peptide of LHA1 (peptide sequence: GLDIETIQQSYT(ph)V). At the end of the reaction, the release of free inorganic phosphate as a measure of phosphatase activity was detected using a colorimetric assay.

As shown in Figure 3.13 A, PLL5 phosphatase domain was able to dephosphorylate LHA1 synthetic phosphopeptide *in vitro*. Its phosphatase activity leads to a significant release of inorganic phosphate, which is not observed upon incubation with the unphosphorylated peptide. Increasing the concentration of phosphopeptide substrate from 50 μM to 300 μM resulted in increasing release of inorganic phosphate (Figure 3.13 B). To confirm that PLL5 phosphatase domain actually de-phosphorylated the phosphorylated residue of the H^+-ATPase peptide, peptides were purified from the reaction mixture at the end of incubation, desalted and analyzed by mass spectrometry (MS). MS analysis confirmed dephosphorylation of the penultimate phospho-Thr residue of H^+-ATPase by PLL5 (Appendix A.7).

A

B

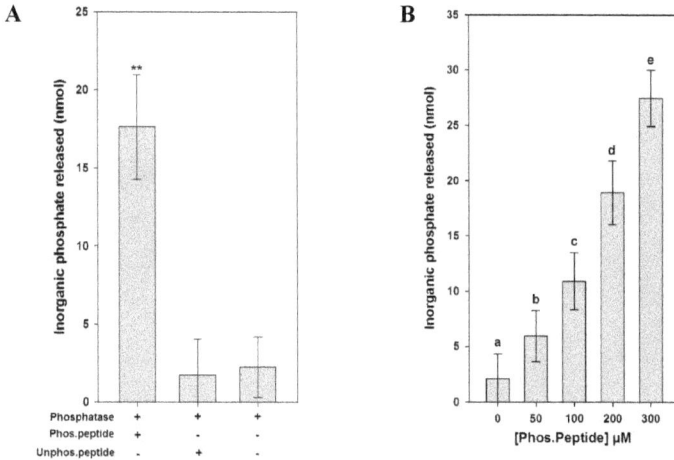

Figure 3.13: PLL5 Phosphatase Domain Can Dephosphorylate LHA1 Phosphopeptide *in vitro*.
(A) One µg of the purified PLL5 phosphatase domain was incubated for 1hr at room temperature with
200 µM of the LHA1 phosphopeptide GLDIETIQQSY(pT)V as substrate. The unphosphorylated
peptide, and an assay without substrate peptide were used as controls. The amount of released inorganic
phosphate was detected by a colorimetric assay (B) Phosphatase activity using increasing concentrations
of LHA1 phosphopeptide as substrate. One µg of purified PLL5 was incubated for 1hr at room
temperature with 50 µM, 100 µM, 200 µM or 300 µM of LHA1 phosphopeptide. (A, B) Activity assays
were carried out using two independent protein isolations in two technical replicates. Graphs show
averages with standard deviations. Small letters indicate significant differences at $p < 0.05$, calculated
by two-tailed t-test. ** represents significant difference at $p < 0.01$ calculated by two-tailed t-test.

To find out if the purified MPK2 WT and CA forms are catalytically active, *in vitro*
kinase assays were performed using myelin-basic protein (MBP) as a standard MAPK
substrate. At the end of the reaction ATP consumption was measured using the Kinase-
Glo® Plus kit (Promega; Fitchburg, WI). As shown in Figure 3.14, both MPK2 forms
showed auto-phosphorylation activity (no-substrate controls). Higher ATP consumption
was detected when they were incubated with MBP, with highest consumption for the
CA form. The data confirm that the CA form has increased activity compared to WT
MPK2.

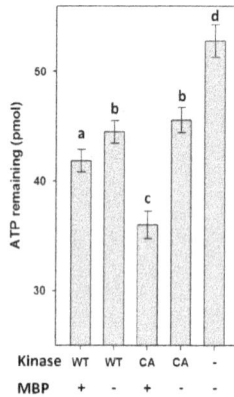

Figure 3.14: MPK2 CA is Catalytically more Active than the WT Form.
MPK2 kinase activity assay using the artificial substrate myelin-basic protein (MBP). Two µg of the purified MPK2 WT or CA forms were incubated 1 hour at 30°C with 6 µg of MBP as a substrate. No-substrate and no-enzyme assays served as controls. The amount of ATP remaining after the reaction was determined by Kinase-Glo® Plus kit (Promega; Fitchburg, WI). Data represent the average of two technical replicates. Equivalent results were obtained using another independent protein isolation. Error bars represent standard deviations. Small letters indicate significant differences at $p < 0.05$ calculated by two-tailed t-test.

To test the ability of MPK2 to phosphorylate the LHA1 peptide (peptide sequence: GLDIETIQQSYTV), the purified MPK2 WT and the CA forms were incubated separately with the synthetic peptide. The reaction was performed as described for MBP substrate above. ATP consumption with the H^+-ATPase C-terminal peptide was significantly higher for the MPK2 CA form in comparison to the no-substrate control. MPK2 WT showed some ATP consumption when incubated with LHA1 peptide but it was less than that of the CA form (Figure 3.15 A), confirming again that the CA form is more active than WT. Increasing the substrate peptide concentration from 10 µM to 200 µM resulted in increasing kinase activity (Figure 3.15 B). To confirm that MPK2 actually phosphorylated the penultimate residue of the H^+-ATPase peptide, peptides were purified after the reaction from the assay mixture and analyzed by MS. MS data confirmed phosphorylation of the penultimate regulatory residue of the H^+-ATPase by recombinant MPK2 *in vitro* (Appendix A.7).

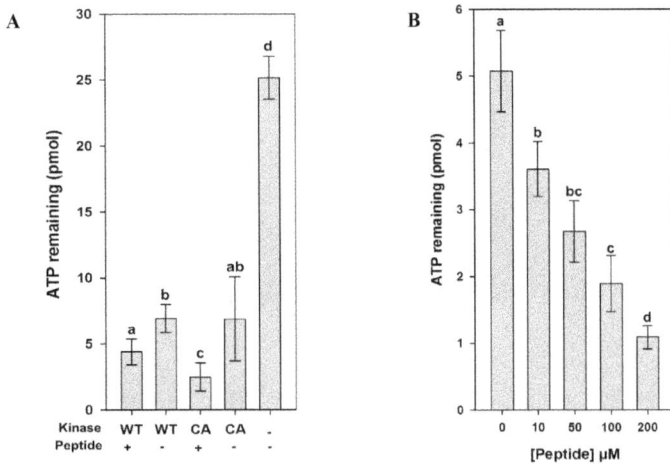

Figure 3.15: MPK2 Can Phosphorylate LHA1 Peptide *in vitro*.
(A) MPK2 kinase activity assay using synthetic LHA1 peptide as substrate. Two µg of the purified MPK2 WT or CA forms were incubated overnight at 30°C with 200 µM of LHA1 peptide GLDIETIQQSYTV as substrate. No-substrate and no-enzyme assays served as controls. The amount of ATP remaining in the reaction was detected by Kinase-Glo® Plus kit (Promega; Fitchburg, WI). (B) Kinase activity using increasing concentrations of substrate LHA1 peptide. Two µg of purified MPK2 CA was incubated with 10 µM, 50 µM, 100 µM or 200 µM of substrate peptide. Activity assays were carried out using two independent protein isolations in two technical replicates. Graphs show averages with standard deviations. The small letters represent significant differences at $p < 0.05$ calculated by two-tailed t-test.

The data confirm the H^+-ATPase as substrate of MPK2 and PLL5. They further suggest that MPK2 and PLL5 may be responsible for phosphorylation and dephosphorylation of the C-terminal regulatory residue of LHA1 upon Systemin perception, respectively. Interestingly, the PLL5 phosphatase phosphopeptide (GLYS(ph)GPLDR) was found in cluster A (rapid dephosphorylation response). The dephosphorylated residue (S160) is located in the regulatory N-terminus of PLL5 (Yu et al., 2003). Under the assumption that similar to PP2A proteins PLL5 phosphatase is activated by dephosphorylation of this residue (Shi 2009; Wera and Hemmings 1995), PLL5 activity would increase upon Systemin treatment. On the other hand, MPK2 was identified in Systemin-induced cluster D with maximum phosphorylation of peptide VTSETDFM(ox)T(ph)EY(ph)VVTR at 15 minutes. The identified phosphopeptide contained the typical MAP-Kinase activating motif (pT)E(pY) (Berriri et al. 2012), suggesting that MPK2 was activated 15 minutes post Systemin perception. Thus, in agreement with data presented in the H^+-ATPase activity assay shown in Figure 3.6 C, PLL5 phosphatase might be involved in dephosphorylation of LHA1 at early Systemin stimulation, which causes inactivation of the H^+-ATPase and medium alkalization, while MPK2 is a candidate kinase for re-phosphorylation at later time points after Systemin treatment, which leads to restoration of H^+-ATPase activity and re-acidification of the medium.

3.2.2. Respiratory Burst Oxidase Homologs (RBOHs)

Another protein with a highly Systemin-specific phosphorylation response was the respiratory burst oxidase. Phosphopeptides of two RBOH isoforms were identified Solyc03g117980 (WHITFLY-INDUCED GP91[phox] [Wfi1], NLSQM(ox)LS(ph)QK) and Solyc06g068680 (Respiratory burst oxidase-like protein, DVFSEPS(ph)QTGR). Upon Systemin stimulation, phosphorylation of these ROS-producing NADPH oxidases occurred at early time points (2 minutes) with a strong transient increase of phosphorylation especially for the Wfi1 phosphopeptide, which was not observed under A17 and water treatments.

The RBOH phosphopeptides identified here were correlated with various Systemin-specific kinases (Figure 3.16) including a receptor-like kinase Solyc09g064270 (r=0.96), a CBL-interacting protein kinase Solyc06g068450 (r=0.91 with Solyc06g068680; r=0.89 with Wfi1) and a MAP triple-kinase Solyc07g055130 (r=0.95 with Wfi1; r=0.93 with Solyc06g068680). Also, an LRR receptor-like kinase (Solyc02g070000) was correlated to both RBOH isoforms (r=0.92 with Wfi1; r=0.95 with Solyc06g068680). This protein is a homolog of the *Arabidopsis* GUARD CELL HYDROGEN PEROXIDE RESISTANT1 (GHR1), which has a critical role in stomatal responses to apoplastic ROS, ABA, high CO_2 concentrations and diurnal light/dark transitions (Hõrak et al. 2016; Hua et al. 2012).

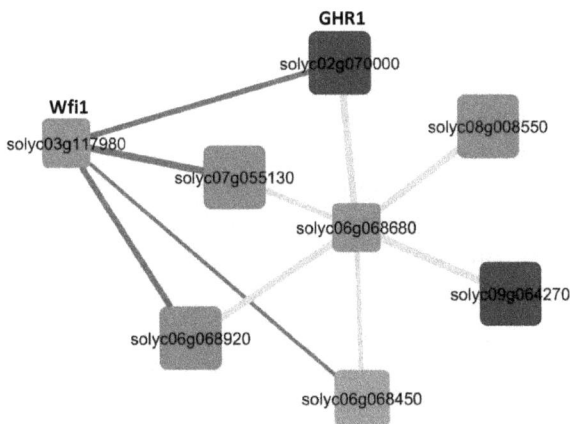

Figure 3.16: Systemin-specific Kinase/Phosphatase Relationships with Respiratory Burst Oxidase Homologs.

A subnetwork is shown of the network for Systemin-specific kinases and phosphatases in Figure 3.9 A and B. Phosphopeptides of RBOHs are shown along with their correlated Systemin-specific kinase/phosphatase partners. Edge color represents the time cluster of the interaction, light gray color indicates earlier clusters (2 min), dark gray indicates later clusters (5 min). Edge width is proportional to the correlation r-value (cutoff r=0.85). Node size indicates the degree in the overall network (Figure 3.9 A & B), Node color represents Mapman function. Brown: respiratory burst oxidases (20.1.1), blue: receptor kinase, pink: calcium-sensing kinases, orange: other kinases of protein.posttranslational modification category (29.4).

In order to test if GHR1 can interact and phosphorylate the peptides of Systemin-induced RBOH isoforms, the intracellular domain of GHR1 (amino acids 624-1024) was expressed in *E. coli* and purified as N-Terminal His_6-tagged protein as described for MPK2 and PLL5. The purified protein was used for *in vitro* activity assays using MBP as substrate. No activity could be detected for purified intracellular domain of GHR1 with MBP as the substrate (data not shown), indicating that the recombinant protein is inactive, or else, that the correct substrate as not been identified. Further experiments will be required to confirm phosphorylation of RBOHs by GHR1 in the future.

3.2.3. Processes at the Cell Wall

Among the target proteins with most significant Systemin-induced phosphorylation responses were cellulose synthase-like proteins. Four phosphopeptides matching cellulose synthase-like proteins (S(ph)SEGDLTLLVDGKPK, Solyc04g077470; SHS(ph)GLMR, Solyc08g076320; SSS(ph)ESGLAELNK, Solyc09g057640; GLIDSQSLSSS(ph)PVK, Solyc09g075550) were identified as being correlated with different Systemin-specific kinases and phosphatases especially during the early Systemin response (Figure 3.17). Systemin-induced phosphorylation changes of cellulose synthases-like proteins were either rapid dephosphorylation (cluster A) or transient phosphorylation at 2 minutes (cluster B). The transient phosphorylation of a receptor-like kinase at 2 minutes (NNTS(ph)SVSPDSVTAK, Solyc12g036330; r=0.99) showed strongest correlation with the transient phosphorylation of peptide S(ph)SEGDLTLLVDGKPK of cellulose synthase-like protein Solyc04g077470. The dephosphorylation of Solyc09g075550 peptide showed strongest correlation with dephosphorylation of peptide GQLPS(ph)GQVVAVK matching an uncharacterized receptor-like cytoplasmic kinase Solyc05g024290 (r=0.99). In addition, a POLTERGEIST-like 1 phosphatase (Solyc08g077150, peptide FVS(ph)PSQSLR) was also identified with high correlation to dephosphorylated peptide GLIDSQSLSSS(ph)PVK (r=0.92).

Figure 3.17: Systemin-specific Kinase/Phosphatase Relationships with Cellulose Synthase-Like Proteins.

A subnetwork is shown of the network for Systemin-specific kinases and phosphatases in Figure 3.9 A and B. Phosphopeptides of cellulose synthase-like proteins are shown along with their correlated Systemin-specific kinase/phosphatase partners. Edge color represents the time cluster of the interaction, light color indicates earlier clusters (2 min), dark color indicates later clusters (15min). Solid edges represent kinase-substrate relationships, dashed edges represent phosphatase-kinase relationships. Node size indicates the degree in the overall network (Figure 3.9 A & B), Node color represents Mapman function. Green: cellulose synthesis proteins (10.2), Blue: receptor kinases, cyan: phosphatases, pink: calcium-sensing kinases, orange: other kinases of protein.posttranslational modification category (29.4), gray: unclassified function. (35).

3.2.4. Vesicle Trafficking and G-proteins Signaling

In total, 52 kinases and 8 phosphatases were found with high correlations to phosphopeptides of vesicle trafficking proteins. While phosphopeptides of adaptins and syntaxins were rapidly dephosphorylated (cluster A), most phosphopeptides derived from the class of vesicle-associated membrane proteins (VAMPs) were transiently phosphorylated at two minutes of Systemin stimulation (cluster B). For example, phosphopeptide NS(ph)DDGYSSDSILR in cluster A matching a KEU-Homolog (Solyc08g079520) showed strong correlation (r=0.99) to phosphopeptide VNS(ph)LVQLPR matching the catalytic subunit of a protein phosphatase 2C (Solyc10g008490). Phosphopeptide VVYIS(ph)PHS(ph)SPGHSEDAFK of VAMP Solyc08g078000 with transient phosphorylation at 2 minutes, on the other hand, showed high correlation with a protein phosphatase Solyc07g053760 (r=0.99) and protein kinase Solyc04g015130 (r=0.99).

Several phosphopeptides of small GTPases were also identified. Most prominent members in this functional group were the ARF GTPase activator Solyc05g023750, and the RAB GTPase activator Solyc12g009610. The phosphopeptide NS(ph)SELGSGLLSR of the ARF-GTPase activator was identified as being rapidly dephosphorylated (cluster A). This peptide correlated highest with phosphopeptide ALIGEGS(ph)YGR of Pti1 kinase Solyc12g098980. In contrast, phosphopeptide TLS(ph)SLELPR of the RAB-GTPase activator Solyc12g009610 was transiently phosphorylated at 2 minutes of Systemin treatment, and this peptide showed highest correlation with one of receptor-like kinase Solyc07g056270. Phosphopeptide AELS(ph)VNR of a second RAB-GTPase (Solyc09g098170) was transiently phosphorylated at 5 minutes, and this peptide correlated highest with the MAP triple kinase Solyc07g055130 and the protein phosphatase 2C Solyc05g055790.

3.2.5. Calcium-related Responses

Several calcium signaling proteins showed high correlation to different Systemin-specific kinases and phosphatases. Two IQ-domain containing proteins (EAS(ph)PKVT(ph)SPR, Solyc06g053450 and FNSLS(ph)PR, Solyc02g087760) were rapidly dephosphorylated upon Systemin stimulation (cluster A). Phosphopeptides of CBL-interacting kinase (Solyc04g076810), protein kinase (Solyc01g108920) and MAP-triple-kinase (Solyc11g033270) as well as a phosphopeptide of phosphatases (Solyc10g008490 and Solyc01g067500) showed strong correlation with dephosphorylation of the IQ-domain containing proteins. A phosphopeptide of calcium-ATPase Solyc02g064680 displayed a transient phosphorylation at 5 minutes (cluster C). Phosphopeptides of phosphatase 2C (Solyc05g055790) and kinases Solyc07g055130 and Solyc01g094920 showed high correlation with the phosphorylation profile of the Ca^{2+}-ATPase.

3.2.6. Transcription Factors

Phosphopeptides of several transcription factor families were over-represented with similar abundance in all Systemin-response clusters (Appendix A.2). The transcription factor with highest degree (37) in the correlation network was Solyc01g009090 a homolog of SERRATE, which was identified by two phosphopeptides S(ph)PPFPPYK and NRS(ph)PPHHPPGR. S(ph)PPFPPYK was transiently dephosphorylated upon Systemin stimulation and showed best correlation to phosphorylation sites of CTR1-like protein kinase 3 (Solyc09g009090), a MAP-triple kinase (Solyc12g088940), and a CBL-interacting kinase (Solyc04g076810). NRS(ph)PPHHPPGR was also dephosphorylated after Systemin treatment. Phosphorylation sites of receptor-like kinases Solyc02g023950 and Solyc09g072810, a CDPK-related kinase Solyc10g078390 and the MAP-triple kinase Solyc11g033270 showed high correlation with that second SERRATE phosphorylation site.

An ARF transcription factor (Solyc05g047460; INS(ph)PSNLR) was rapidly dephosphorylated upon Systemin stimulation. The peptide of this protein showed very high correlation with a CBL-interacting protein kinase homolog (Solyc04g076810) and the CTR1-like protein kinase 3 (Solyc09g009090). A WRKY transcription factor (Solyc02g080890; EES(ph)PESESWAPNK) was identified 15 min after Systemin stimulation (cluster D). This phosphopeptide showed high correlation to Solyc08g068250 and Solyc04g015130 protein kinases and Solyc07g053760 PP2C phosphatase.

3.2.7. Interactions with other Phytohormone Signaling Pathways

Among the specifically Systemin-induced phosphorylation/dephosphorylation events were several phosphopeptides of proteins known from other hormone signaling pathways. For example, phosphopeptides of abscisic acid response factors (ASSESSSEDEADS(ph)AEEAGSSK, Solyc04g081340; VGS(ph)PGNDFIAR, Solyc01g108080), auxin response factor (INS(ph)PSNLR, Solyc05g047460), the auxin efflux transport protein PIN4 (PSNFEENCAPGGLVQS(ph)SPR, Solyc05g008060) and ET receptor EIN4 (GGT(ph)PELVDTR, Solyc05g055070) were rapidly dephosphorylated upon Systemin treatment (cluster A). Transient phosphorylation at 2 minutes of Systemin treatment was observed for another phosphorylation sites in the ABA-response factor Solyc04g081340 (RSS(ph)ASATQVQAEEQAPR), and for SSS(ph)RLNLSTR in the ET synthesis enzyme ACS (Solyc03g007070). A third phosphopeptide of EIN4 (GGS(ph)QTDSSISTSHFGGK) was transiently phosphorylated at 5 minutes of Systemin treatment. Kinases with high correlations to the phosphorylation patterns in cluster A were CTR1-like protein kinase 3 (Solyc09g009090) a negative regulator of ET signaling, LRR1 receptor kinase homolog (Solyc11g011020) and a CBL-interacting protein kinase (Solyc04g076810).

3.2.8. Receptor-like Kinases Induced Upon Systemin Stimulation

Ninety-eight phosphopeptides belonging to 60 receptor-like kinases (RLKs) were identified in the whole data set (listed in Appendix A.8). Several of these phosphopeptides were rapidly dephosphorylated upon Systemin treatment (cluster A), while others showed transient phosphorylation after 2, 5 and 15 minutes post Systemin stimulation (clusters B, C and D). About one third of them were classified as Systemin non-responsive (cluster F). Some RLKs were identified with several phosphopeptides that had different phosphorylation profiles and therefore, were clustered to different clusters. For example, GHR1 (Solyc02g070000) was identified with 5 phosphopeptides. Three of them were rapidly dephosphorylated (MAAITGFS(ph)PSK; M(ox)AAITGFS(ph)PSK and LDVKS(ph)PDR) upon Systemin stimulation, while QATSNPS(ph)GFSSR was transiently phosphorylated after 2 minutes and GSSEIIS(ph)PDEK was classified as Systemin non-responsive (cluster F).

To find out which RLKs are specifically induced by Systemin and not by other elicitors, it was checked whether the *Arabidopsis thaliana* homologs of the receptors in Appendix A.8 were found in previous studies to be phosphorylated in response to the bacterial elicitor Flg22, the DAMP oligogalacturonides (OGs), or the fungal effector xylanase (Benschop et al. 2007; Mattei et al. 2016; Nühse et al. 2007; Rayapuram et al. 2014). This comparison showed that 17 RLKs were also (de)phosphorylated in response to other elicitors and, therefore, not Systemin specific. In addition, about half of the phosphopeptides identified for these RLKs belonged to the Systemin non-responsive cluster (cluster F). These RLKs were excluded from further analysis.

Among the remaining 43 Systemin-specific RLKs, 20 phosphopeptides of 12 RLKs showed high normalized intensity after 2 and 5 minutes of Systemin treatment and were induced later or not at all after A17 stimulation, indicating that their activity might be essential for the Systemin signaling pathway. It has been reported that in addition to post-translational regulation by (de)phosphorylation, RLKs may also be regulated at the transcriptional level, like for example FLS2 and PORK1 that are transcriptionally upregulated after Flg22 perception and wounding, respectively (Mersmann et al. 2010; Xu et al. 2018). Therefore, in order to get further evidence for a potential involvement of any of these 12 RLKs in Systemin perception and signal transduction, the transcript levels of these RLKs was tested by qRT-PCR after wounding. In addition to wild-type tomato plants, transgenic tomato plants were used in this experiment that overexpressed either Prosystemin, or a mutated version of Prosystemin (D-2A/D18A), in which the Aspartate residues flanking the Systemin cleavage sites were substituted by Alanine (Beloshistov et al. 2018; Dreizler 2018). Tomato plants overexpressing Prosystemin were reported to have a constitutively upregulated wound response, while plants expressing the D-2A/D18A mutated version of Prosystemin are behaving like wild type, presumably because Prosystemin cannot be processed (Beloshistov et al. 2018; McGurl et al. 1994). For this experiment, leaf tissue was collected before (0 hr, control) and 1 and 2 hours after wounding. cDNA synthesized from total leaf RNA was used as template in qRT-PCR. Figure 3.18 shows the expression of the RLKs transcripts after wounding relative to unwounded wild-type control plants.

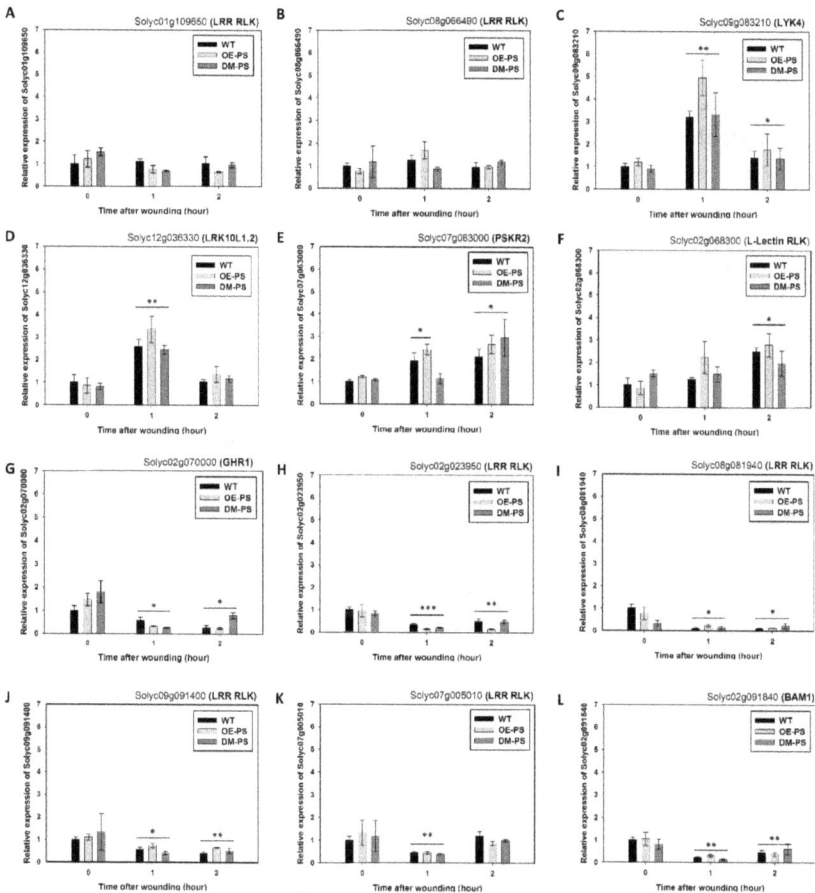

Figure 3.18: Expression of Systemin-Specific RLKs in Wounded Tomato Plants.
Relative expression of **(A)** Solyc01g109650 (LRR-RLK), **(B)** Solyc08g066490 (LRR-RLK), **(C)** Solyc09g083210 (LYK4 homolog), **(D)** Solyc12g036330 (LRK10L1.2), **(E)** Solyc07g063000 (PSKR2), **(F)** Solyc02g068300 (L-Lectin RLK), **(G)** Solyc02g070000 (GHR1), **(H)** Solyc02g023950 (LRR-RLK), **(I)** Solyc08g081940 (LRR-RLK), **(J)** Solyc09g091400 (LRR-RLK), **(K)** Solyc07g005010 (LRR-RLK), and **(L)** Solyc02g091840 (BAM1) at 0, 1 and 2 hours after wounding in wild type (WT) and transgenic tomato plants overexpressing Prosystemin (PS-OE) or double-mutated Prosystemin

(DM-PS) relative to unwounded wild-type plants. Expression level were normalized to the level of EF (elongation factor 1α) expression. qRT-PCR was performed on cDNA synthesized from total RNA from pooled leaf material of six 2-week-old plants. The data represent means of two technical replicates with standard deviation. The expression of each RLKs at 1 or 2 hours after wounding is shown relative to that of unwounded control plants (0 hour). *: $p \leq 0.05$; **: $p \leq 0.01$; ***: $p \leq 0.001$ (two-tailed t-test).

It was expected that the RLKs involved in the Systemin signaling pathway may be upregulated at the transcript level after wounding. This was the case for only 2 of the 12 tested RLKs (Figure 3.18 C & D), which were the LYSIN MOTIF (LysM) CONTAINING RLK4 homolog (LYK4; Solyc09g083210) and a receptor homologous to AtLRK10L1.2 (Solyc12g036330). The expression of the PHYTOSULFOKINE RECEPTOR2 (PSKR2; Soly07g063000) was also transiently upregulated after wounding (Figure 3.18 E), although its phosphopeptide was classified in the Systemin non-responsive cluster F. This may imply a different role for LYK4 and Solyc12g036330 in the Systemin signaling pathway as compared to PSKR2.

The LYK4 homolog belongs to LysM domain-containing RLKs that are well known to bind various types of peptidoglycans and chitin (Hohmann et al., 2017). AtLYK4 was found to be involved in chitin signaling and has an overlapping function with AtLYK5 (Cao et al. 2014; Wan et al. 2012). This was demonstrated in Atlyk4/Atlyk5 double mutant that showed a complete loss of chitin response in comparison to single mutants (Cao et al. 2014). Chitin recognition by LysM containing RLKs enables plants to recognize symbiotic bacteria or sense and induce resistance against pathogenic fungi (Buist et al. 2008; Cao et al., 2014).

The Arabidopsis homolog of Solyc12g036330 is the Leaf Rust 10 Disease-Resistance Locus Receptor-Like Kinase1.2 (LRK10L1.2; Lim et al. 2014). AtLRK10L1.2 was found to be involved in ABA signaling and to act as a positive regulator for drought tolerance via the induction of stomatal closing (Lim et al. 2014). Subcellular localization of this receptor to the PM was shown to be critical for its function in ABA signaling and drought tolerance (Lim et al. 2014; Shin et al. 2015).

PSKR2 is a LRR-RLK that is involved in the perception of the tyrosine-sulfated pentapeptide Phytosulfokine (PSK; Tabata and Sawa 2014). PSK is a plant growth factor, which promotes plant development via enhanced cell expansion and division (Sauter 2015; Song et al. 2017). Recently, it was found that PSK also is involved in plant immune responses against certain pathogens (Sauter 2015; Zhang et al. 2018b). In Arabidopsis as well as in tomato there are two receptors that can bind and perceive PSK

(Hartmann et al. 2015; Zhang et al. 2018b). In tomato, PSKR1 rather than PSKR2 functions as the major PSK receptor in immune responses (Zhang et al. 2018b). In contrast to Systemin, PSK perception induces H^+-ATPase activation causing PM hyperpolarization, apoplastic acidification, and as a result cell wall extension (Ladwig et al. 2015).

The findings of Wang et al. (2018) and Xu et al. (2018) suggest that more than one RLK is essential for proper Systemin perception. To find out if the RLKs filtered out from this work are involved and important in the Systemin signaling pathway, a loss-of function study was performed in *S. peruvianum* cell cultures as well as in tomato plants. This aimed to test the effect of the candidate RLKs as well as SYR1, SYR2 and PORK1 knock-outs (KO) on Systemin-triggered early and late events.

The list of the candidate-RLKs is shown in Table 3.1. The selection criteria were: (1) the RLK is (de)phosphorylated only in response to Systemin and not in response to MAMP elicitors, (2) it is (de)phosphorylated later or not at all in response to A17, (3) it is transiently (de)phosphorylated with rapid kinetics (2 or 5 minutes) upon Systemin stimulation, and (4) it is upregulated at the transcript level after wounding. All four criteria were fulfilled by only two RLKs; the homologs of *At*LYK4 and *At*LRK10L1.2. Another three candidates were also included in this list, since they missed only one of these criteria. They included an uncharacterized LRR-RLK (Solyc01g109650), whose expression level did not change after wounding (Figure 3.18 A) but had the highest transient phosphorylation for some of its identified phosphopeptides after 5 minutes of Systemin treatment. Also included was GHR1 which was downregulated at transcript level upon wounding (Figure 3.18 G) but was identified with several phosphopeptides one of which with a high transient phosphorylation 2 minutes after Systemin treatment. In addition, GHR1 was highly correlated to the Systemin-responsive RBOHs (Figure 3.16). PSKR2 was also considered in the candidate list, since its expression was upregulated after wounding, although its phosphopeptide was classified as not systemin-responsive (cluster F).

Table 3.1: List of Candidates for Systemin-Responsive RLKs.
Listed are the RLKs accessions, type of ectodomain, functional description, accession of *Arabidopsis* homologs, wound response, sequence of identified phosphopeptides, number of spectra for each phosphopeptide, Systemin and A17 clusters for each peptide as well as A17 response shift.

Accession (abbreviation)	Ecto-Domain	Description	Arabidopsis homolog	Wound response	Phosphopeptide	Spectra	Sys cluster	A17 cluster	A17 shift
Solyc09g083210 (LYK4)	LysMII	LysM containing receptor-like kinase 4 (LYK4) homolog	AT2G23770	Up-regulated	SIDLFTDVSEEGLS(ph)PR	24	C	F	later
					SLYLGS(ph)K	29	B	E	later
Solyc12g036330 (LRK10L1.2)	WAK/LRK10L1	Leaf rust 10 Disease-resistance locus receptor-like protein kinase-like 1.2 homolog	AT1G18390	Up-regulated	SYAGSS(ph)LITR	35	B	E	later
Solyc07g063000 (PSKR2)	LRRXb	Phytosulfokine receptor 2 (PSKR2)	AT5G53890	Up-regulated	S(ph)SDTFVPSK	37	F	E	earlier
Solyc01g109650 (LRRXIV)	LRRXIV	Receptor like kinase, RLK	AT2G16250	No change	LEGTSS(ph)LK	22	C	E	later
					LEGTS(ph)SLK	13	B	F	later
					SSS(ph)DVAAVPAAASAHK	29	B	A	earlier
Solyc02g070000 (GHR1)	LRRIII	Guard cell hydrogen peroxide-resistant1 (GHR1) homolog	AT4G20940	Down-regulated	QATSNPS(ph)GFSSR	10	B	F	later
					GSSEIIS(ph)PDEK	39	F	B	earlier
					M(ox)AAITGFS(ph)PSK	45	A	E	later
					LDVKS(ph)PDR	60	A	A	equal
					MAAITGFS(ph)PSK	21	A	E	later

The genes for candidate RLKs listed in Table 3.1, as well as SYR1, SYR2 and PORK1, were mutated using the CRISPR/Cas9 genome editing method. To introduce mutations at the selected target sites of each of the candidate genes, a CRISPR/Cas9 construct with two sgRNAs was designed and cloned according to Xing et al. (2014) . CRISPR/Cas9 constructs were stably introduced into *S. peruvianum* cells by particle bombardment (Cedzich et al. 2009), and into tomato plants by Agrobacterium-mediated transformation (Bosch et al. 2014b).

3.3.   RLK Loss of Function Analysis

3.3.1.   Characterization of *S. peruvianum* RLK Knocked-out Cell Cultures

Two to three different cell suspension cultures were established from three independent transgenic calli for each RLK CRISPR/Cas9 construct.

The cell suspension cultures were genotyped to identify the type of mutations they harbored and to exclude cultures that still contained wild-type alleles of the corresponding RLK. Thus, gene fragments spanning the mutation target sites in each gene were amplified by PCR and cloned into pCR2.1 TOPO®. At least eight different clones were sequenced for each target site. In each cell suspension culture several different mutations were found at each target site. Most of them were missense mutations leading to truncated proteins as a result of early stop codons. All cell suspension cultures for which the wild-type sequence was found in at least one of the eight clones was excluded from further analysis. Appendix A.9 lists the mutations identified in the RLKs of all cell suspension cultures used for further analysis.

3.3.2.   Response of *S. peruvianum* RLK loss-of-function (KO) Cell Cultures to Systemin

Cell suspension cultures that were confirmed to be mutated for any given RLK were tested for their alkalization response to 10 nM Systemin and compared to that of wild type. To avoid false negative results because of any general defect in the cell's ability to alkalize the growth medium, the alkalization response of the mutated cell cultures to the Flg22 peptide was tested and used as positive control. If only Systemin perception is impaired in the mutated cell cultures, it is expected that these cells will show a Flg22 response similar to that of wild type. The average ΔpH (maximum change in pH after stimulation) was plotted against time and compared to wild type (Figure 3.19).

It can be noticed that the Flg22 alkalization response of all RLK KO cell suspension cultures was not significantly different from that of the wild-type culture (Figure 3.19, black bars) indicating that there was no general defect in the ability to induce an alkalization response, by e.g. insufficient cell density, or unrelated mutations. In contrast, the Systemin response of all but one RLKs KO cell cultures was significantly lower than that of wild type, with the single exception of the *syr2* KO culture, which showed a wild-type response to Systemin. ΔpH was strongly reduced in *lrk10l1.2*, *pskr2* and *pork1* KO lines, while the SYR1 knock-out caused complete abolishment of the Systemin response (Figure 3.19, gray bars).

The data confirm SYR1 as the major Systemin receptor. In addition, the LRK10L1.2 receptor (Solyc12g036330), PORK1 and PSKR2 appear to play important roles in this early part of the signaling cascade.

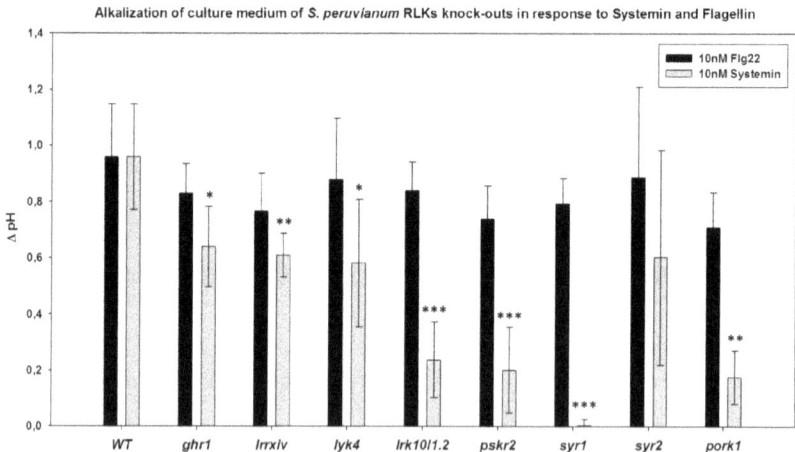

Figure 3.19: The Alkalization Response of RLKs KO Cell Suspension Cultures after Systemin and Flg22 Stimulation.
Ten ml of 7-day old mutated cell suspension cultures as well as wild type were stimulated with 10 nM Systemin. In parallel, another 10 ml aliquot of the same cells batches was stimulated with 10 nM Flg22. The pH change was monitored continuously over time. ΔpH (the difference in pH at 0 min just before addition of the peptide and 8 min when the maximum was reached) is shown for different cell cultures and peptide treatments. The data represents the mean of at least 2 independent mutant cell suspension cultures, each tested at least 3 times independently. Error bars represent standard deviation. Significant difference between the wild type and each of the RLK KO cultures are indicated as *: $p < 0.05$, **: $p < 0.01$, or ***: $p < 0.001$ (two-tailed t-test).

3.3.3. Characterization of RLK Loss of Function in Gene-Edited Tomato Plants

The CRISPR/Cas9 constructs designed to knock out the candidate *RLK* genes were introduced into tomato plants (*S. lycopersicum*, cv. UC82b) by Agrobacterium-mediated transformation.

T_1 plants were genotyped by PCR for the presence of the transgene (the CRISPR/Cas9 genome editing construct) and for any mutations in the targeted *RLK* genes. Different mutations were detected at both guide-RNA target sites for most *RLK* genes except for *PSKR2* and *SYR2*, for which mutations were found only at the first target site (the one 5' of the other; Appendix A.10). All detected mutations were missense mutations leading to truncated proteins because of early stop codons. Based on the genotyping results T_1 plants were selected that are homozygous for the mutation and lack the transgene (the CRISPR/Cas9 genome editing construct). These plants were kept in the greenhouse for seed collection.

3.3.4. Wound Response of RLK Knock-out Tomato Plants

The ability of RLK KO plants to respond to wounding was analyzed in homozygous mutants of the T_2 generation. GHR1 is involved in stomatal closure in response to different stimuli in *Arabidopsis* (Hua et al. 2012). Therefore, stomatal movement was analyzed in two *ghr1* T_2 lines in response to wounding. It was found that the stomatal aperture index of these mutants is higher than that of wild type after wounding (data not shown), which indicated that stomatal movement is impaired in these mutants, consistent with the function of GHR1 in *Arabidopsis*. It also was noticed that stomata are smaller, and that stomatal density is higher in *ghr1* mutants as compared to wild type (data not shown). These preliminary results indicated the possible involvement of *Sl*GHR1 in wound-induced stomatal closure. However, these results need to be confirmed by further analysis.

The induction of *Proteinase Inhibitor II* (*PI-II*) expression was analyzed by qPCR as a wound response marker in *ghr1* and all other RLK KO mutants. PI-II is a well-established target of systemin-mediated wound signaling, which reaches its maximum of expression about 6-8 hours post wounding (Peña-Cortés et al. 1995; Ryan 2000). The wounding experiment was performed as described in section 3.2.8, except that the wounded and distal leaves were collected at 8 hours after wounding. qRT-PCR was performed to detect the level of the *PI-II* transcript in RLK KO plants relative to unwounded wild-type plants (Figure 3.20). In this experiment, 2 independent KO lines were tested for LYK4, LRRXIV, GHR1 and SYR2. Only one line was tested for SYR1, PSKR2 and LRK10L1.2, since their transformation is still ongoing at the time of writing of this thesis. PORK1 could not be included in this experiment, since its transformation was started later than that of the other RLKs and is still running.

Figure 3.20 (A) shows that the local wound response of most RLK mutants is similar to that of wild type. Only in *pskr2* mutants, induction in *PI-II* expression was about three times higher than in the wild type. However, this difference was not statistically significant. In contrast, differences to the wild type were observed for the systemic wound response, which was most obvious for the *lrk10l1.2* mutant (Figure 3.20 A). There was a significant reduction in the systemic induction of *PI-II* expression in *lrk10l1.2*. *pskr2* mutants showed a higher induction of *PI-II* expression also in the systemic tissue, which again was insignificant. Knocking out SYR1 and SYR2 had no effect on the wound response marker gene expression, which is consistent with data reported by Wang et al. (2018).

It can be noticed that the most of the RLK KO plants have somewhat higher *PI-II* expression than the wild type before wounding (Figure 3.20 B). This difference was statistically significant for one of the *lyk4* and the *ghr1* lines. Apparently, there is constitutive upregulation of *PI-II* expression in these mutants under normal conditions. Despite elevated levels of *PI-II* transcript in the unwounded state, *PI-II* expression was still wound-inducible in these plants (Figure 3.20 A).

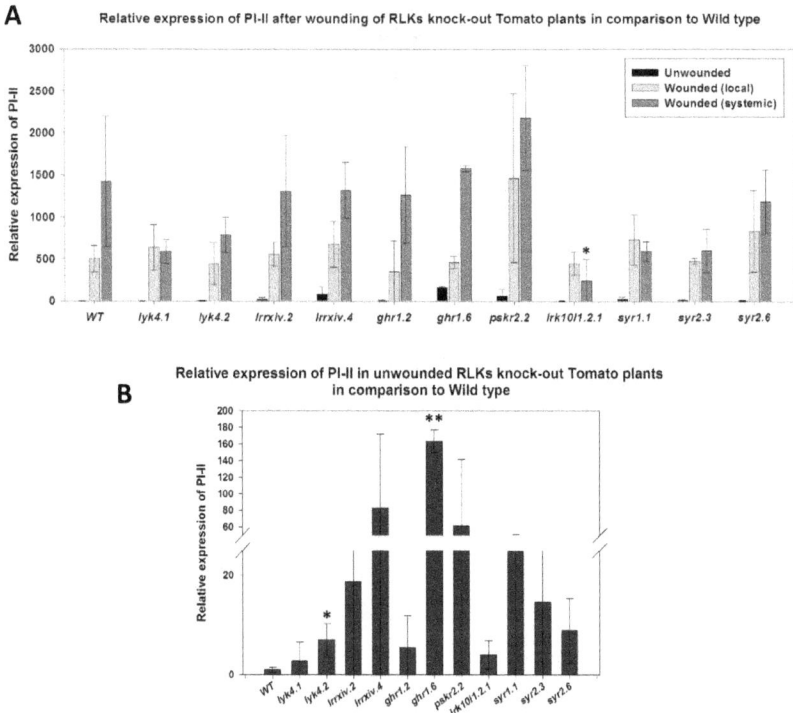

A Relative expression of PI-II after wounding of RLKs knock-out Tomato plants in comparison to Wild type

B Relative expression of PI-II in unwounded RLKs knock-out Tomato plants in comparison to Wild type

Figure 3.20: *PI-II* Expression after Wounding in RLK Mutants in Comparison to Wild Type.
(A) Local and systemic *PI-II* expression level in wounded RLK KO plants relative to unwounded wild-type plants quantified by qRT-PCR. (B) A subgraph from the bar graph in A showing the constitutive *PI-II* expression in unwounded RLK mutants in comparison to wild type. Plants of two independent mutated lines for LYK4, LRRXIV, SYR2 and GHR1 were tested, while only one line was tested for PSKR2, LRK10L1.2 and SYR1. qRT-PCR was performed on cDNA synthesized from total RNA of pooled leaf material from two 2-week-old T_2 plants. *PI-II* expression level was normalized to the level of EF1α, Actin and Ubiquitin expression. Data represent the mean of 3-4 biological replicates each tested in duplicate with standard deviation. The relative expression of *PI-II* locally and systemically after wounding was always compared to that of unwounded wild-type plants. Significant differences are indicated by *: $p < 0.05$, **: $p < 0.01$ (two-tailed t-test).

To summarize the results obtained for the alkalization and wound responses of the RLK KOs, it can be concluded that the function of SYR1 as the Systemin receptor is more relevant for early than for late events of Systemin signaling. On the other hand, PORK1 appears to be involved in early events of the Systemin cascade (alklinization; this work) as well as in late Systemin responses (defense gene expression, insect resistance; Xu et al. (2018)). Likewise, the LRK10L1.2 receptor appears to contribute to both, early and late events of the Systemin signaling pathway. Interestingly, knock out of PSKR2 negatively affected the early events of Systemin perception and positively affected the late events of the pathway suggesting that its role in Systemin signaling is in a way different from that of PORK1 and LRK10L1.2.

4. Discussion

Despite all biotic stresses exerted on plants by pathogens, herbivore attacks and wounding, the world is still green. This is because plants, despite their vulnerable nature, are not just defenseless bystanders. They have an army of different defense mechanisms, which work together in harmony to detect and ward off dangerous attacks. Likewise, tomato, as one of the economically important crop plants, has the same defense mechanisms with an extra crucial player "Systemin".

During this work the Systemin-induced intracellular events as well as Systemin-specific kinases and phosphatases were studied using a time-course large-scale phosphoproteomic approach.

Phosphoproteomics was successfully used before to identify many differentially phosphorylated proteins in response to PTI and ETI elicitors. Benschop et al. (2007), Nühse et al. (2007) and Rayapuram et al. (2014) used different quantitative phosphoproteomics approaches to identify phosphorylation events occurring in *Arabidopsis* in response to the bacterial elicitor Flg22. Moreover, Benschop and his colleagues examined the phosphorylation events exerted in *Arabidopsis* in response to the fungal effector xylanase. Mattei et al. (2016) analyzed early changes in the *Arabidopsis* membrane phosphoproteome in response to the DAMP elicitor OGs. Recently, Kadota et al. (2019) employed a phosphoproteomic screen to identify the differential phosphorylation events induced by the bacterial effector avrRpt2 in *Arabidopsis* seedlings.

In this work, the Systemin-induced early phosphorylation events were studied using *S. peruvianum* cell suspension culture. This cell suspension culture was used over many years as a model to study early Systemin responses because of its high sensitivity to the peptide; indeed sub-nanomolar concentration of Systemin is sufficient to trigger a response (Felix and Boller 1995; Meindl et al. 1998). It also has, like other plant cell suspension cultures, the advantage of allowing the study of sequential events with a good temporal resolution, avoiding the problems associated with the analysis of a multi-factorial plant composed of multiple tissue and cell types exposed to diverse signals (Boller and Felix 2009; Hampp et al. 2012). In fact, the use of such cell cultures was helpful to Benschop et al. (2007) and Nühse et al. (2007) to unravel the early phosphorylation events induced by the plant immunity elicitors in *Arabidopsis*.

Like other elicitor-induced intracellular immune signaling (Macho and Zipfel 2014), Systemin-induced early phosphorylation events, mediated by cell surface-localized receptor kinase(s), are expected to occur at the PM (Schaller and Oecking 1999; Scheer and Ryan 1999; Wang et al. 2018). Due to low abundance of membrane proteins and the deep embedding of some of them in the PM, these early receptor-mediated signaling events are difficult to detect in crude cell lysate (Orsburn et al. 2011). As a consequence, microsomal membrane proteins were used frequently for phosphoproteomics analysis of signaling pathways (Mattei et al. 2016; Nühse et al. 2004; Orsburn et al. 2011), which were used in this work as well.

4.1. The Use of the Inactive Analog A17 to Define Systemin-specific Responses

Antagonistic, mutated or chemically modified peptides were used as tools for the characterization of peptide-receptor interaction and to unravel the unknown functions of peptides in plants (Czyzewicz et al. 2015; Schaller 1998). In cases of genetic redundancy or if mutant plants are unavailable, the dominant-negative effect of antagonistic peptides may result in a loss-of-function phenotype, which was helpful in identifying the function of some redundant *Arabidopsis* CLE peptides (Song et al. 2013; Xu et al. 2015). Antagonistic peptides bind competitively to native peptide receptors but are unable to activate the signaling cascade if functionally critical amino acids are missing (Czyzewicz et al. 2015). Keinath et al. (2010) used the antagonistic mutated peptide Flg22Δ2 as control to analyze the changes in detergent-resistant membranes (DRMs) proteome in response to Flg22 stimulation in *Arabidopsis*.

The inactive A17 peptide has an antagonistic effect on Systemin signal transduction as shown in Figure 3.1, and as described previously by Pearce et al. (1993) and Felix and Boller (1995). The use of this inactive peptide as negative control in this work enabled the identification of Systemin-specific phosphopeptides, which comprised approximately half of the total phosphopeptides identified from the whole experiment (Figure 3.8). Including of A17 analog as a control thus gave an insight into Systemin-specific signaling processes.

4.2. Systemin Specifically Induced Intracellular Events, Kinases and Phosphatases

In this work, k-means clustering identified dynamic and transient changes of different phosphorylation events induced by Systemin (Figure 3.5), including transient phosphorylation events of 56 kinases and 17 phosphatases. Pairwise Pearson correlation analysis was used to identify their potential substrates in the Systemin-specific cellular processes (Figure 3.9 and Appendix A.6). Correlation analysis was previously used to reconstruct the topology of phosphorylation networks to characterize the regulatory interactions between phosphoproteins in response to changes in nutrient and organic-solute supply (Duan et al. 2013). Within the phosphorylation network PM proteins were identified as a top layer for signal induction, while cytosolic and nuclear proteins were considered as an effector layer. In budding yeast, correlation analysis enabled the identification of candidate proteins involved in the transition from budding to filamentous growth under stress conditions (Zhang et al. 2013). It is worth noting that Pearson correlation-based inference methods result in undirected networks (Duan et al. 2013). Directionality assignments can be performed either by further statistical analysis like calculating the ratio of partial variance between each two nodes connected by an edge (Duan et al. 2013), or by building protein-protein correlation networks with the help of up-to-date databases (Jorgensen and Linding 2008; Lin et al. 2015), which does not eliminate the need of further experiments to prove or disprove the predicted interactions.

Functional categories of signaling, transport, cellular and cell wall-associated processes were significantly over-represented in Systemin-specific proteins. These functional categories were over-represented also in previous studies of elicitor-induced signaling pathways (Benschop et al. 2007; Keinath et al. 2010; Mattei et al. 2016) indicating, as previously observed, that innate immune signaling pathways converge on common intracellular defense responses (Boller and Felix 2009; Choi and Klessig 2016).

Phosphorylation and dephosphorylation responses observed at 2 minutes after treatment (clusters A and B) represent early Systemin-induced responses. These early responses included (de)phosphorylation of e.g. MAP triple kinases, Ca^{2+} sensing proteins (CBL-interacting kinases [CIPKs] and Calcium-dependent Protein Kinases [CDPKs]), and other proteins belonging to the functional categories vesicle trafficking, transport, and signaling. Among them were many Systemin-specific RLKs with phosphorylation patterns peaking after 2 or 5 minutes. In addition, Systemin stimulation exerted very rapid transient changes in phosphorylation profiles of PM-located proteins that are known to belong to the Systemin signaling cascade (Sagi et al. 2004; Schaller and Oecking 1999), including two proton pumps that were temporarily dephosphorylated after Systemin treatment, and two respiratory burst oxidases (RBOHs) isoforms that showed a strong increase of phosphorylation after 2 minutes. Later responses of the Systemin signaling cascade included proteins that showed a phosphorylation peak after 15 minutes of stimulation such as some MAPKs, G-proteins and transcription factors.

Among the early Systemin-induced responses was the ET signaling pathway represented by the phosphorylation of ACS, the enzyme responsible for ET precursor biosynthesis, already at 2 minutes. EIN4 as one of ET receptors showed a transient phosphorylation peak after 5 minutes upon Systemin stimulation. On the other hand, none of the enzymes involved in JA biosynthesis were identified in the whole data set. This is in consistent with the observation that the *S. peruvianum* cell culture used in this study, does not express wound and jasmonate response marker proteins in response to Systemin stimulation (Felix and Boller 1995; Schaller and Oecking 1999). The apparent inability to activate the expression of late response markers may be due to the toxicity of some of these proteins (e.g. polyphenol oxidase), which may have resulted in the selection of mutations causing a downstream block in the signaling pathway (Schaller and Oecking 1999).

Proteins involved in synthesis, signal transduction and response to other phytohormones such as ABA, auxin, cytokinin and SA showed different phosphorylation patterns at different time points. Modulation of signaling pathways of these hormones upon Systemin stimulation reflects the crosstalk between them and their involvement in balancing growth with responses to environmental cues, which is required for the establishment of appropriate and effective defense responses in plants (Denancé et al. 2013; Pieterse et al. 2012).

Proteins involved in cell wall processes were mostly either rapidly dephosphorylated or phosphorylated 2 or 5 minutes upon Systemin stimulation. Such processes are required for cell wall reinforcement as one of the common PTI-associated defense responses (Boller and Felix 2009; Houston et al. 2016). Recently, in a metabolic profiling analysis of tomato plants overexpressing Prosystemin or treated with Systemin, it was shown that Systemin enhanced accumulation of lignans (Pastor et al. 2018), which known as antimicrobial defense metabolites that may also contribute to cell wall reinforcement and hard wood durability (Wang et al. 2013).

4.2.1. Systemin-induced Modulation of the PM H$^+$-ATPase (LHA1)

Among the remarkable Systemin-induced phosphorylation events were those exerted on the penultimate activating C-terminal sites of two PM H$^+$-ATPases; LHA1 and AtAHA1 homolog. The phosphorylation profiles of these phosphopeptides as well as the *in vitro* ATPase assay (Figure 3.6) showed that PM proton pumps were deactivated rapidly upon Systemin stimulation, then, after 15 minutes their activity was increased and restored. In *Arabidopsis*, Flg22 perception causes dephosphorylation of the penultimate C-terminal residue of AHA1 and 2 as well (Benschop et al. 2007; Nühse et al. 2007).

Pearson correlation analysis showed that LHA1, is correlated to 12 different kinases and two phosphatases (Figure 3.10) among which are MPK2 and a homolog of PLL5, which belongs to Clade C of PP2C phosphatases. *In vitro* assays demonstrated that they can phosphorylate and dephosphorylate the penultimate residue (T955) of LHA1, respectively (Figures 3.13 and 3.15).

Several phosphatases were reported to be involved in regulating H$^+$-ATPase activity. For example, protein phosphatase 2A (PP2A-A) was shown to interact with the C-terminal regulatory domain of *Arabidopsis* AHA2 (Fuglsang et al. 2006). PP2C-D phosphatases were demonstrated to negatively regulate the activity of the same proton pump during auxin-induced cell expansion (Spartz et al. 2014). In addition, phosphatases play a crucial role in regulating innate immune responses. Recently, a protein phosphatase PP2C38 was identified as a negative regulator of Botrytis-Induced Kinase1 (BIK1), a direct substrate of some PRR complexes (Couto et al. 2016). Rice XB15, a PP2C phosphatase was found to associate with the PRR XA21 attenuating innate immune responses against the bacterial pathogen *Xanthomonas oryzae* pv. *oryzae* (Park et al. 2008). Similarly, EFR, which is phylogenetically closely related to XA21 and binds the bacterial elf18 peptide, is associated with PLL4 and PLL5 (XB15 orthologs), which negatively regulate EFR-mediated responses against *Pseudomonas syringae* pv. tomato DC3000 (Holton et al. 2015).

Systemin stimulation induced rapid dephosphorylation of Ser160 located in the regulatory N-terminus of the tomato PLL5 homolog (Yu et al., 2003). Little is known about the exact mechanism of PP2C activity regulation (Shi 2009). However, it is reported that some phosphatases, such as PP2A phosphatases and the human Pyruvate Dehydrogenase Phosphatase1 (PDP1), a PP2C phosphatase that dephosphorylates pyruvate dehydrogenase to promote respiratory metabolism, can be activated by dephosphorylation and inactivated by phosphorylation (Bradshaw et al. 2017; Martin et al. 2010; Shin et al. 2014; Wera and Hemmings 1995). Assuming similar regulation for the tomato PLL5 homolog, its rapid dephosphorylation upon Systemin stimulation most likely induces its activity allowing it to rapidly dephosphorylate the penultimate C-terminal residue of LHA1, inactivating the proton pump and leading to apoplast alkalization.

MPK1 and 2 were reported to be involved in the tolerance against abiotic stresses such as oxidative and heat stress, mechanical wounding and UV-B radiation in tomato (Higgins et al. 2007; Holley et al. 2003; Nie et al. 2013). They were reported also to be activated in response to Systemin, several oligosaccharide elicitors such as chitin and plant cell wall polygalacturonates, and ETI (Higgins et al. 2007; Holley et al. 2003; Pedley and Martin 2004). This again confirms that innate immune signaling pathways converge on common intracellular events (Boller and Felix 2009; Higgins et al. 2007; Holley et al. 2003). These two MAPKs are 95% identical at the amino acid sequence level (Holley et al. 2003; Kandoth et al. 2007). Specific silencing of either MPK1 or MPK2 resulted in the same reduction in expression of proteinase inhibitors and other defense genes in response to Systemin as co-silencing of both together (Kandoth et al. 2007). Co-silencing of MPK1 and MPK2 also abolished Prosystemin-mediated resistance to herbivory exerted by *Manduca sexta* (Kandoth et al. 2007). MPK1 and MPK2 were shown to be involved in Systemin signaling upstream of JA biosynthesis (Kandoth et al. 2007).

MPK1/2 activity was detected upon 5 minutes of Systemin stimulation with maximum increase after 15 minutes (Higgins et al. 2007). Similarly, the MAP-kinase activating motif (pT)E(pY)-containing phosphopeptide (Berriri et al. 2012) of MPK2 was identified in this work with increasing phosphorylation starting 5 minutes upon Systemin stimulation and with a maximum peak after 15 minutes. The data suggest that MPK1/2 is activated to re-phosphorylate LHA1 to restore its activity later after Systemin stimulation. This is consistent with the presented *in vitro* kinase assay results (Figure 3.15) showing MPK2-mediated phosphorylation of the penultimate regulatory threonine of LHA1. Phosphorylation of this residue is likely to reactivate the proton pump for re-acidification of the extracellular space dampening the initial alkalization response induced by Systemin. It is worth noting that the amino acid sequence of the identified MPK2 phosphopeptide is completely identical in MPK1, which implicates that MPK1 may also be involved in the Systemin signaling pathway. However, the *in vitro* kinase assay was performed only with recombinant MPK2, not with MPK1.

4.2.2. Systemin-induced Modulation of RBOHs

An *At*RBOHD homolog, Wfi1, is one of two Systemin-induced RBOHs, whose silencing was shown previously to compromise wound-induced systemic expression of *PI-II* in tomato (Sagi et al. 2004). Wfi1 was identified with several phosphopeptides, one of which showed a phosphorylation peak lasting from 2 to 5 minutes after Systemin stimulation. The phosphorylated residue (S340) in this phosphopeptide is conserved in *At*RBOHD (S347). In *Arabidopsis* this residue is known to be phosphorylated by BIK1 and Ca^{2+}-sensing proteins such as CDPKs in response to Flg22, the bacterial elongation factor Tu (elf18), and chitin perception (Dubiella et al. 2013; Kadota et al. 2014; Kadota et al. 2015; Kimura et al. 2017; Ogasawara et al. 2008; Zhang et al. 2010). ROS production by RBOHs also is directly regulated by Ca^{2+} binding to the EF-hand motifs within their N-terminal cytosolic region (Kärkönen and Kuchitsu 2015; Ogasawara et al. 2008). Furthermore, in some signaling cascades it was found that small G-proteins, can directly interact with and phosphorylate RBOHs (Kimura et al. 2017).

Tomato Protein Kinase 1b (TPK1b; Solyc06g005500) is the tomato homolog of BIK1 that regulates defense responses to chewing insects and necrotrophic fungi (AbuQamar et al. 2008). It was recently reported that this cytoplasmic receptor-like protein kinase (Sakamoto et al. 2012) is the substrate of PORK1, one of the RLKs crucial for Systemin perception (Xu et al. 2018). Silencing of TPK1b impaired *PI-II* expression upon Systemin application in tomato (Xu et al. 2018).

In the current work, TPK1b was identified with a phosphopeptide that was classified to be Systemin non-responsive (cluster F), which explains why it was not included in Systemin-specific kinases and phosphatases correlation network. A closer look at the phosphorylation profile of this phosphopeptide shows that it has a slight phosphorylation maximum 5 minutes after Systemin stimulation. On the other hand, Figure 3.16 shows that a CIPK (Solyc06g068450), which has a phosphorylation peak 2 minutes upon Systemin treatment, is correlated to both Systemin-induced RBOHs. Under the assumption that Wfi1 might be regulated in the same way as *At*RBOHD (Kadota et al. 2014) it would be of great interest to test whether TPK1b along with the CIPK, as a Ca^{2+} sensing protein, is responsible for the phosphorylation of Wfi1 at S340 upon Systemin perception.

4.3. RLKs Involved in Systemin Perception

The recently identified PM located LRR Systemin receptor SYR1 as well as its closely related homologs SYR2 and PORK1 (Wang et al. 2018; Xu et al. 2018) were not identified among the 3312 phosphopeptides from this work. Technical limitations could be a reason for this; some RLKs might be strongly imbedded in the cell membrane preventing them from being enriched by the extraction protocol. This has been observed for the Flg22 receptor FLS2 in previous phosphoproteomics studies using Flg22-stimulated *Arabidopsis* cell suspension cultures (Benschop et al. 2007; Nühse et al. 2007). Another reason could be that the first sampling time point (2 minutes) is already too late for detecting auto-phosphorylation of RLKs involved in Systemin perception. Flg22 perception, for example, induces very rapid phosphorylation of FLS2 and BAK1 within seconds of stimulation (Schulze et al. 2010).

During this work it was demonstrated that a knock-out of SYR1 in the *Solanum peruvianum* cell culture completely abolished the alkalization response induced by nanomolar concentrations of Systemin (Figure 3.19). This observation confirms SYR1 as a high-affinity genuine Systemin receptor, which is responsible for Systemin perception and signal transduction including proton pump inactivation. On the other hand, loss of SYR1 function in tomato plants did not have any effect on *PI-II* expression after mechanical wounding (Figure 3.20). Similarly, the induction of *PI-II* expression after mechanical wounding was the same in the tomato introgression line IL3-3 lacking the genomic region of SYR1, and in SYR1-complemented IL3-3 (Wang et al. 2018). These observations can possibly be explained by the activity of DAMPs like OGs that are released by mechanical wounding or, alternatively, by an additional Systemin-independent pathway for *PI-II* induction (León et al. 2001; Savatin et al. 2014). Interestingly, sensitivity to systemin was increased by *SYR1* expression in IL3-3 plants, resulting in increased *PI-I* expression upon Systemin application, and in increased resistance to herbivorous insects (Wang et al. 2018). These data confirm that Systemin-induced *PI-I* expression is exerted through perception of Systemin by SYR1, which is important also for the resistance against herbivore insects (Wang et al. 2018).

In contrast, knock out of the SYR1-related receptor SYR2 did not have any effect on the Systemin-induced alkalization response and wound-induced *PI-II* expression (Figure 3.19 and 3.20). SYR2 was shown to bind to Systemin with very low affinity and it was suggested that it might have a role in the perception of a different, perhaps Systemin-related, ligand (Wang et al. 2018).

PORK1 was recently reported to regulate Systemin, wound, and immune responses in tomato (Xu et al. 2018). After wounding and Systemin application, PORK1 transcript levels are increased, and its knock down rendered the plants more susceptible to the tobacco hornworm *Manduca sexta* (Xu et al. 2018). In contrast to SYR1, PORK1 function was found to be important for wound-induced *PI-II* accumulation (Xu et al. 2018). The Systemin-induced alkalization response of *pork1* mutated cell suspension cultures was significantly weaker than that of wild-type cells indicating that the PM proton pump was not fully deactivated upon Systemin perception (Figure 3.19). Altogether, the data indicate that PORK1 is important for proper Systemin signal transduction, including proton pump inactivation, and its presence is more critical for wound-induced defense responses than SYR1.

It is proposed by Xu et al. (2018) that Systemin perception may by carried out through multiple RLKs including SYR1, SYR2 and PORK1, like it is the case for the fungal MAMP chitin, which is perceived through multiple LysM RLKs with different affinities for chitin (Cao et al. 2014; Petutschnig et al. 2010). However, the specific contribution of these receptors to Systemin perception and their interaction need to be tested by further experiments.

4.4. PSKR2 is Important for Systemin Signal Transduction and Phytosulfokine Attenuates the Wound-Induced Defense Response

Addressing a possible involvement of additional RLKs in Systemin perception and signal transduction, a group of RLKs was selected that showed remarkable induction of phosphorylation upon Systemin stimulation. In a first experiment it was tested whether these RLKs are regulated by mechanical wounding at the transcript level (Figure 3.18). Three of them were transiently upregulated after wounding (Figure 3.18 C, D and E), and most of them showed a higher level of expression in transgenic tomato plants overexpressing Prosystemin (Figure 3.18). This agrees with the previously observed constitutive upregulation of wound-inducible genes in Prosystemin overexpressing plants (McGurl et al. 1994).

One of the identified RLKs in the data set was PSKR2, i.e. one of the two phytosulfokine (PSK) receptors (Matsubayashi et al. 2002; Stührwohldt et al. 2011).

PSK is a disulfated pentapeptide that was initially identified as a growth factor that promotes growth of low-density cell suspension cultures (Matsubayashi and Sakagami 1996). It is derived from precursors encoded by small gene families, which display differential expression throughout the plant life cycle (Lorbiecke and Sauter 2002; Sauter 2015; Srivastava et al. 2008).

In addition to its growth-promoting function (Motose et al. 2009), PSK also is involved in plant innate immune responses (Matsubayashi 2014). The expression of some PSK precursor genes is induced upon pathogen infection, elicitor treatment or wounding (Loivamäki et al. 2010; Matsubayashi et al. 2006), and resistance to necrotrophic pathogens, such as *Alternaria brassicicola* in *Arabidopsis* and *Botrytis cinerea* in tomato, is reduced when PSK signaling is impaired (Igarashi et al. 2012; Mosher et al. 2013; Zhang et al. 2018b). Resistance against biotrophic pathogens such as *P. syringae*, on the other hand, is increased in plants impaired in PSK signaling (Igarashi et al. 2012; Mosher et al. 2013; Zhang et al. 2018b). Mosher et al. (2013) showed that PSK signaling mutants inoculated with *P. syringae* accumulate elevated levels of salicylate and SA-responsive *PR*-gene transcripts, whereas the JA-responsive genes *PDF1.2* and *OPR3* were repressed. Enhanced SA signaling correlates with the observed increased resistance to biotrophic pathogens, and the decreased JA levels are consistent with increased susceptibility to necrotrophic pathogens (Mosher and Kemmerling 2013). In addition, these mutants exhibit early senescence, a SA-associated response, and are impaired in wound healing, a JA-associated response (Amano et al. 2007; Loivamäki et al. 2010; Matsubayashi et al. 2006). Consequently, a link between phytosulfokine signaling and the regulation of SA/JA homeostasis during immunity responses was suggested (Amano et al. 2007; Matsubayashi et al. 2006). It is well known that SA-dependent plant defense directed against biotrophic pathogens works antagonistically to JA-dependent defense against necrotrophic pathogens (Spoel and Dong 2008). Therefore, it was proposed that PSK signaling shifts the hormone homeostasis in favor of the JA pathway and negatively regulates SA accumulation (Mosher et al. 2013; Sauter 2015).

In tomato, it was found that PSK has the same binding affinity (K_d) to both of its receptors PSKR1(K_d = 6.36 µM) and PSKR2 (K_d = 7.21 µM) (Zhang et al. 2018b). However, the association rate constant (K_a) for the PSK-PSKR1 interaction is higher than that with PSKR2 (K_a for PSKR1= 2.68 x 10^3 M^{-1}s^{-1}, for PSKR2= 9.86 x 10^2 M^{-1} s^{-}

[1]) (Zhang et al. 2018b). In Arabidopsis, cell expansion, hypocotyl and root elongation are regulated positively by both receptors in a partially redundant manner, with PSKR1 playing the dominant role (Amano et al. 2007; Hartmann et al. 2015; Stührwohldt et al. 2011). Likewise, the immune-modulatory function of PSK also is mainly exerted through PSKR1 (Igarashi et al. 2012; Mosher et al. 2013; Zhang et al. 2018b). In contrast, medium acidification in *S. peruvianum* cell cultures in response to PSK depended on PSKR2 (Bachelor thesis of Cecile Landenberger, 2018). In this work, it was shown that knocking out PSKR2 in the *S. peruvianum* cell culture and in tomato plants affected Systemin and wound-induced responses (Figures 3.19 and 3.20). CRISPR/Cas9-mediated receptor knock out was achieved with a construct comprising two PSKR2-specific sgRNAs. Off-target mutations in *PSKR1* can be excluded, since the sequence of the two sgRNAs had mismatches at their 3'-end just before the Protospacer Adjacent Motif (PAM). It is well known that mismatches proximal to the PAM are not tolerated and impair cleavage activity of the Cas9 endonuclease (Zheng et al. 2017). Off-target mutations were also not observed for the SYR1 sgRNA in SYR2 and vice versa, for which similar mismatches existed proximal to the PAM sequence.

Both PSK receptors have a LRR ectodomain with a PSK-binding island domain, and a cytosolic kinase domain, with overlapping Calmodulin (CaM) binding site and Guanylate Cyclase (GC) center, responsible for cyclic GMP (cGMP) production (Hartmann et al. 2015; Kwezi et al. 2011; Zhang et al. 2018b). It was reported in *Arabidopsis* that PSKR1 can physically interact with the H^+-ATPase proton pumps and with BAK1, which both in turn interact with the Cyclic Nucleotide-Gated Channel17 (CNGC17) that is regulated by cGMP and CaM to control Ca^{2+} influx (Ladwig *et al.*, 2015). A recent report suggested that PSKR1, with its GC activity, might act as a scaffold protein that brings key enzymes of signaling cascades in the correct place to ensure ordered spatial and temporal stimulus-specific message generation (Irving et al. 2018). Such scaffold function was demonstrated also for the RLK Feronia, which was shown to promote ligand-induced complex formation between FLS2 or EFR and their co-receptor BAK1 (Stegmann et al. 2017). This complex formation was inhibited through RALF23 peptide perception by Feronia (Stegmann et al. 2017).

Since PSKR2 has same structure as PSKR1 (Hartmann et al. 2015), it is possible that PSKR2 also acts as a scaffold protein that facilitates the interaction of signal transduction core elements. Such a scenario could explain the observed reduction of alkalization response in *pskr2*-mutated cell cultures upon Systemin stimulation (Figure 3.19). Along these lines it can further be speculated that PSKR2 is required for bringing the Systemin receptor, its co-receptor, if there is any, and the proton pump together to allow proper and complete inactivation of the proton pump through PLL5, the suggested phosphatase that inactivates LHA1 upon Systemin perception. The proposed scenario could be verified by *in vivo* interaction experiments such as BiFC assay. Whether the assembly of this interaction complex is induced by PSK or not will be an interesting question for further experiments.

The PSKR2 knock out in tomato seemed to cause an increase in local and systemic *PI-II* expression after mechanical wounding, albeit this increase was statistically not significant (Figure 3.20). This observation suggests that PSK signaling may negatively regulate *PI-II* expression induced upon wounding. This proposition is consistent with what was already published by Motose et al. (2009) who showed that PSK significantly reduced the accumulation of stress-related gene transcripts including the proteinase inhibitors upon wounding. These authors suggested that PSK is involved in the attenuation of stress response to keep a healthy balance between growth and defense and avoids excessive non-specific activation of immune responses (Igarashi et al. 2012; Mosher et al. 2013; Motose et al. 2009). To confirm the involvement of PSK in attenuating wound-induced *PI-II* expression, PSK can be applied to wounded wild-type tomato plants followed by *PI-II* quantification.

4.5. Involvment of ABA in Systemin and Wound Signaling Pathway through LRK10L1.2

Among the wound-upregulated receptors (Figure 3.18) was a homolog of *At*LRK10L1.2. Two of the identified phosphopeptides of this RLK showed a strong phosphorylation maximum 2 minutes after Systemin stimulation. Recently, *At*LRK10L1.2 was reported to be one of the receptors involved in ABA-induced stress responses for drought resistance (Lim et al. 2014; Shin et al. 2015). *Atlrk10l1.2* mutants are ABA insensitive, hypersensitive to drought, and impaired in stomatal closure in response to ABA (Lim et al. 2014). However, LRK10L1.2 does not affect the expression of ABA response marker genes suggesting that ABA signaling through this receptor works independently of the well-established PYR/RCAR-PP2C-SnRK2 pathway for ABA signaling (Lim et al. 2014). How this receptor is involved in ABA stress responses is unclear (Lim et al. 2014).

Among the phytohormones that accumulate locally and systemically in tomato plants upon mechanical wounding or Systemin application besides JA is the stress hormone ABA (León et al. 2001). ABA is capable of inducing local and systemic expression of wound-defense genes such as *PI-II* (Chao et al. 1999; León et al. 2001; Peña-Cortés et al. 1995), while ABA-deficient tomato plants do not respond to wounding or Systemin treatment (Chao et al. 1999; Peña-Cortés et al. 1995). Application of ABA restored the response of these plants to Systemin and wounding, which indicates that the Systemin/wound signaling pathway requires ABA for induction of JA and PI-II accumulation (Chao et al. 1999; Peña-Cortés et al. 1995). Interestingly, the response to ABA is impaired when JA biosynthesis is inhibited, and restored upon JA application (Peña-Cortés et al. 1995). The data suggest that in the Systemin and wound signaling pathway the site of JA action is located downstream of ABA (Peña-Cortés et al. 1995).

The accumulation of ABA at the wound site may result from local dehydration, and may help to prevent further water loss by inducing stomatal closure, or pathogenic infection at the site of injury by the induction of defense responses (León et al. 2001; Mittler and Blumwald 2015; Savatin et al. 2014; Takahashi and Shinozaki 2019).

It is well known that physiological concentrations of ABA cause dephosphorylation of the penultimate activating C-terminal threonine residue of the PM H^+-ATPase in both suspension cell cultures (Chen et al. 2010) and guard cells (Hayashi et al. 2011). ABA-mediated inactivation of the proton pump contributes to depolarization of the cell membrane resulting in stomatal closure (Falhof et al. 2016; Mittler and Blumwald 2015). However, the mechanisms of ABA-mediated dephosphorylation and inactivation of the PM H^+-ATPase is not known (Falhof et al. 2016).

During this work it was shown that knock out of *SlLRK10L1.2* impaired the Systemin-induced alkalization response (Figure 3.19). Since *Atlrk10l.2* mutants are deficient in ABA responses, a role for ABA in the Systemin-induced dephosphorylation of LHA1 is proposed, and this effect of ABA may be mediated by LRK10L1.2. This hypothesis can be tested through treating of *Sllrk10l1.2* mutated cell suspensions with ABA and comparing their alkalization response to that of wild-type cells.

Since ABA was previously shown to accumulate both locally at the wound site as well as systemically (Peña-Cortés et al. 1995), impaired ABA signaling in *Sllrk10l1.2* mutants would be expected to affect the induction of *PI-II* expression in local and systemic tissues. However, wound-induced expression of *PI-II* in the *Sllrk10l1.2* mutant was significantly reduced only in the systemic leaves (Figure 3.20). This can possibly be explained by LRK10L1.2 being required only for ABA-mediated systemic PI-II accumulation, while local *PI-II* expression is mediated by an ABA/LRK10L1.2-independent mechanism. Alternatively, the wound-induced systemic signal might be impaired in *Sllrk10l1.2* mutants. Using grafting experiments performed with different tomato mutants it was demonstrated that the systemic wound signal requires biosynthesis of JA at the site of injury and the ability to perceive it in distal tissues (Schilmiller and Howe 2005). Similar to ABA-deficient tomato plants that are not able to accumulate JA in response to Systemin application and wounding (Peña-Cortés et al. 1995), impaired ABA signaling in *Sllrk10l1.2* may affect the local induction of JA biosynthesis and, consequently, impair systemic signaling. This can be tested using reciprocal grafting of *Sllrk10l1.2* mutants with wild-type tomato plants by wounding the rootstock and testing *PI-II* mRNA accumulation in the scion.

It is worth noting, that Flg22 treatment causes dephosphorylation of a phosphopeptide located at the C-terminus of *At*LRK10L1.2 (Benschop et al. 2007). In an analysis of the OG response, this peptide was detected only in mock-treated samples (Mattei et al. 2016). In the present work, the corresponding residue in the tomato LRK10L1.2 homolog was found to be also dephosphorylated upon Systemin stimulation. In addition, Systemin stimulation also caused the rapid (2 min) phosphorylation of a second residue in the same region, as well as in the juxtamembrane (JM) domain. Since other well-studied RLKs are known to be activated by phosphorylation at the JM and C-terminus (Wang et al. 2005, 2008), it is proposed that Systemin perception induces the activation of LRK10L1.2. This seems to be a specific Systemin response, since *lrk10l1.2*-mutated cell suspension cultures showed wild-type alkalization response after Flg22 treatment (Figure 3.19).

4.6. Systemin Triggers Stomatal Closure through GHR1 Regulation

One of the RLKs correlated to both Systemin-induced RBOH isoforms is GHR1. This RLK was identified with four phosphopeptides one of which was highly phosphorylated 2 minutes after Systemin stimulation.

In *Arabidopsis*, GHR1 is involved in stomatal closure in response to apoplastic ROS, ABA, high CO_2 concentrations, light stress and diurnal light/dark transitions (Devireddy et al. 2018; Hõrak et al. 2016; Hua et al. 2012; Sierla et al. 2018). In the *Atghr1* mutant stomatal movement in response to different stimuli such as MeJA, SA and Flg22 is impaired (Hua et al. 2012). It was shown that GHR1 acts downstream of ROS and upstream of Ca^{2+} to activate the anion channel SLAC1 required for stomatal closure (Devireddy et al. 2018; Hua et al. 2012). Its activity is negatively regulated by the protein phosphatase ABA-Insensitive2 (ABI2) and the protein kinase High Leaf Temperature1 (HT1) during ABA- and CO_2-induced signaling (Hõrak et al. 2016; Hua et al. 2012).

In vitro kinase assays showed that HT1 can directly phosphorylate several residues of *At*GHR1, some of them at the JM domain (Hõrak et al. 2016; Sierla et al. 2018). All *Sl*GHR1 phosphorylation sites identified after Systemin treatment were in the JM domain, two of which were homologous to HT1 phosphorylation sites of *At*GHR1. One of these phosphorylation sites was rapidly dephosphorylated and the other one showed a high phosphorylation peak 2 minutes after Systemin treatment. This suggests that *Sl*GHR1 is probably phosphorylated by HT1 along with other kinases in response to Systemin stimulation.

In 2012 Hua et al. showed that *At*GHR1 can phosphorylate the anion efflux channel SLAC1. In contrast, Sierla et al. (2018) reported very recently that *At*GHR1 is an inactive pseudokinase, which may activate SLAC1 as a scaffolding protein mediating the interaction with CDPK3. A closer look at the ATP binding site of *Sl*GHR1 revealed that it lacks some residues indispensable for kinase activity (Figure 4.1). *Sl*GHR1 thus seems to be an inactive pseudokinase, just like *At*GHR1 (Murphy et al. 2014; Sierla et al. 2018). Consistent with this conclusion, no activity was detected for the recombinant intracellular domain of *Sl*GHR1 in *in-vitro* kinase assay using MBP as substrate (data not shown).

The similarity to *At*GHR1 suggests that *Sl*GHR1 may be involved in stomata regulation, rather than the wound response. Indeed, knock out of this receptor in *S. peruvianum* cell suspension culture and in tomato plants showed that it is not required for Systemin-induced medium alkalization or *PI-II* expression (Figure 3.19 and 3.20). In contrast, a role in stomata regulation was supported by preliminary phenotyping experiments. Stomatal closure in response to wounding was impaired in *Slghr1* mutants compared to wild type. Furthermore, stomata seemed to be smaller in size but higher in density in *Slghr1* compared to wild type (preliminary data, not shown). In *Atghr1*, on the other hand, stomatal density was not changed (Hua et al. 2012).

Taking all these observations into consideration, it can be hypothesized that GHR1 may be involved in stomatal closure as one of the defense responses triggered by the Systemin-induced apoplastic ROS release (Hua et al. 2012; Sierla et al. 2018), rather than directly phosphorylating RBOH proteins as might be inferred from Figure 3.16.

Confirmation of this hypothesis and the apparent function of *Sl*GHR1 in stomatal development and regulation in tomato will require more detailed analyses in the future.

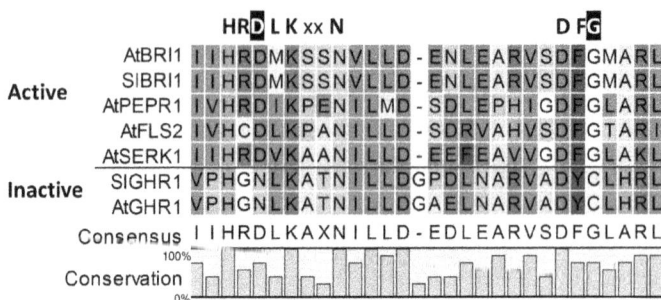

Figure 4.1: *Sl*GHR1 ATP-binding lacks residues indispensable for kinase activity.
A multiple amino acid sequence alignment of subdomains VIb and VII of the catalytic core of the kinase domains of active RLKs and the inactive *At*GHR1 and *Sl*GHR1. Residues highlighted in black are considered indispensable for kinase activity. Bold residues are highly conserved in most active kinases. The alignment was performed using CLC Main Workbench v. 8.1, Qiagen.

4.7. LYK4 and the Uncharacterized LRRXIV RLK are Not Involved in Systemin Signaling Pathway

LYK4 is a LysM-domain receptor for chitin binding and perception (Petutschnig et al. 2010), conferring resistance against bacterial and fungal pathogens in *Arabidopsis* (Wan et al. 2012). In this work, a homolog of this receptor was identified with a phosphopeptide that showed the highest phosphorylation maximum of all identified RLKs at 2 minutes of Systemin stimulation. Its transcript level was increased upon wounding as well (Figure 3.18). Despite that, knock out of this receptor did affect neither the Systemin-induced alkalization in *S. peruvianum* cell suspension culture, nor wound-induced *PI-II* expression in tomato plants. The same was observed for the LRRXIV receptor, which showed the highest phosphorylation maximum 5 minutes post Systemin treatment. Therefore, these receptors are not involved in immediate Systemin responses. However, they may still contribute to Systemin-induced defense priming.

Defense priming is a status in which the plant upon stimulus perception undergoes changes at the physiological, transcriptional, metabolic, and epigenetic levels allowing it to effectively initiate a faster and/or more aggressive defense response against future challenges, which results in increased resistance and/or stress tolerance (Frost et al.

2008; Mauch-Mani et al. 2017). For example, it was found that the experience of wounding stress primes *Arabidopsis* plants resulting in enhanced resistance to subsequent infection by the pathogenic fungus *B. cinerea* by inducing faster defense responses (Beneloujaephajri et al. 2013; Chassot et al. 2008). Likewise, overexpression of Prosystemin in tomato (Xu et al. 2018) and *Arabidopsis* (Zhang et al. 2018a) leads to enhanced resistance to *B. cinerea*.

Defense priming can be mediated by transcriptional reprogramming, for example by increasing the transcription of defense-related RLKs (Mauch-Mani et al. 2017; Schenk et al. 2014). In *Arabidopsis*, treatment with Flg22 or OGs induces the expression of many RLKs, among which are *LRK10L1.2* and *LYK4* (Denoux et al. 2008). The same was reported in Prosystemin-overexpressing tomato plants, where *FLS2* was one of many overexpressed genes (Bubici et al. 2017). Alternatively, priming may be achieved by phosphorylation of RLKs as described by Benschop et al. (2007), who showed that Flg22 perception triggered the phosphorylation of different RLKs. In the present work, Systemin perception induced phosphorylation of several RLKs, among which were LYK4 homolog and the uncharacterized LRRXIV RLK. However, confirmation of a role for these kinases in Systemin-induced defense priming requires further experimentation. For example, the efficiency of wounding-induced priming of resistance against fungal and/or bacterial infections should be compared in *lyk4*-mutated and wild-type plants. Alternatively, a direct role of LYK4 in chitin perception can be tested by comparing the chitin-induced alkalization response of *lyk4*-mutated and wild-type cell suspension cultures.

4.8.   Conclusion and Outlook

In the presented work, Systemin was shown to induce a series of cellular events that ranged from PM to cytosolic and nuclear signaling. In addition, 56 Systemin-specific kinases and 17 Systemin-specific phosphatases were identified. Putative substrates of Systemin-specific kinases and phosphatases were identified by pairwise correlations of their Systemin-induced phosphorylation time profiles. These substrates are likely to include novel players in the Systemin signaling pathway. Two of these players, MPK2 and PLL5, were confirmed as the kinase and the phosphatase involved in controlling the PM H^+-ATPase (LHA1) activity in response to Systemin. *In vivo* assays like BiFC assay could verify the interaction of MPK2 and PLL5 with LHA1. In addition, knocking out

PLL5 in *S. peruvianum* cell suspension culture and testing its alkalization response to Systemin will confirm whether it is the phosphatase responsible for inactivating LHA1 in this pathway. Furthermore, since MPK1 was reported to be involved in the Systemin signaling pathway (Kandoth et al. 2007), it will be interesting to test if it is also involved in the same way as its close relative MPK2.

During this work SYR1 was confirmed as the main receptor responsible for Systemin perception, as demonstrated by the compromised Systemin-induced alkalization response of *syr1*-mutated cell cultures. As already reported (Wang et al. 2018), SYR1 was found not to be required for *PI-II* expression after wounding. PORK1 was shown to be involved in Systemin signal transduction (this study) and wound signaling (Xu et al. 2018). Testing the binding affinity of PORK1 ectodomain to Systemin might give further insight into the role of PORK1 in a Systemin receptor complex, as suggested by Xu et al. (2018).

Knock out of some Systemin-specific RLKs gave insight into the crosstalk of Systemin with other phytohormones such as ABA and PSK, represented by LRK10L1.2 and PSKR2, respectively.

A possible role for PSKR2 in Systemin signal transduction was suggested by the reduced Systemin-induced alkalization response of *pskr2*-mutated cell cultures. Furthermore, PSKR2 might be involved in PSK-mediated attenuation of wound-induced defense responses. Further experiments are required to confirm these observations. Additionally, it will be of great interest to test whether PSKR1, as the main PSK receptor involved in plant immunity (Zhang et al. 2018b), is involved in the Systemin signaling pathway as well.

To get further insight into the specific roles of the individual receptors in Systemin signaling and the wound response, detailed analyses of ROS and Ca^{2+} accumulation as well as MAPK activity will be required in *syr1*, *pork1*, *lrk10l1.2* and *pskr2* mutated cell suspension cultures and tomato mutants.

5. Bibliography

AbuQamar, S., M. F. Chai, H. Luo, F. Song, and T. Mengiste. 2008. "Tomato Protein Kinase 1b Mediates Signaling of Plant Responses to Necrotrophic Fungi and Insect Herbivory." *THE PLANT CELL* 20(7):1964–83.

Acevedo, Flor E., Loren J. Rivera-Vega, Seung Ho Chung, Swayamjit Ray, and Gary W. Felton. 2015. "Cues from Chewing Insects — the Intersection of DAMPs, HAMPs, MAMPs and Effectors." *Current Opinion in Plant Biology* 26:80–86.

Adeboye, Peter Temitope, Maurizio Bettiga, and Lisbeth Olsson. 2014. "The Chemical Nature of Phenolic Compounds Determines Their Toxicity and Induces Distinct Physiological Responses in Saccharomyces Cerevisiae in Lignocellulose Hydrolysates." *AMB Express* 4(1):1–10.

Alborn, H. T., T. V. Hansen, T. H. Jones, D. C. Bennett, J. H. Tumlinson, E. A. Schmelz, and P. E. A. Teal. 2007. "Disulfooxy Fatty Acids from the American Bird Grasshopper Schistocerca Americana, Elicitors of Plant Volatiles." *Proceedings of the National Academy of Sciences* 104(32):12976–81.

Alborn, H. T., T. C. J. Turlings, T. H. Jones, G. Stenhagen, J. H. Loughrin, and J. H. Tumlinson. 1997. "An Elicitor of Plant Volatiles from Beet Armyworm Oral Secretion." *Science* 276(5314):945–49.

Amano, Y., H. Tsubouchi, H. Shinohara, M. Ogawa, and Y. Matsubayashi. 2007. "Tyrosine-Sulfated Glycopeptide Involved in Cellular Proliferation and Expansion in Arabidopsis." *Proceedings of the National Academy of Sciences* 104(46):18333–38.

Andolfo, Giuseppe and Maria R. Ercolano. 2015. "Plant Innate Immunity Multicomponent Model." *Frontiers in Plant Science* 6:Article 987.

Ardito, Fatima, Michele Giuliani, Donatella Perrone, Giuseppe Troiano, and Lorenzo Lo Muzio. 2017. "The Crucial Role of Protein Phosphorylation in Cell Signalingand Its Use as Targeted Therapy (Review)." *International Journal of Molecular Medicine* 40(2):271–80.

Arimura, G. and I. S. Pearse. 2017. "From the Lab Bench to the Forest: Ecology and Defence Mechanisms of Volatile-Mediated 'Talking Trees.'" *Advances in Botanical Research* 82:3–17.

Bartels, Sebastian and Thomas Boller. 2015. "Quo Vadis, Pep? Plant Elicitor Peptides at the Crossroads of Immunity, Stress, and Development." *Journal of Experimental Botany* 66(17):5183–93.

Bastías, Daniel A., M. Alejandra Martínez-Ghersa, Jonathan A. Newman, Stuart D. Card, Wade J. Mace, and Pedro E. Gundel. 2018. "Jasmonic Acid Regulation of the Anti-Herbivory Mechanism Conferred by Fungal Endophytes in Grasses" edited by D. Gibson. *Journal of Ecology* 106(6):2365–79.

Basu, Saumik, Suresh Varsani, and Joe Louis. 2018. "Altering Plant Defenses: Herbivore-Associated Molecular Patterns and Effector Arsenal of Chewing Herbivores." *Molecular Plant-Microbe Interactions* 31(1):13–21.

Baydoun, E. A. H. and S. C. Fry. 1985. "The Immobility of Pectic Substances in Injured Tomato Leaves and Its Bearing on the Identity of the Wound Hormone." *Planta* 165(2):269–76.

Beloshistov, Roman E., Konrad Dreizler, Raisa A. Galiullina, Alexander I. Tuzhikov, Marina V. Serebryakova, Sven Reichardt, Jane Shaw, Michael E. Taliansky, Jens Pfannstiel, Nina V. Chichkova, Annick Stintzi, Andreas Schaller, and Andrey B. Vartapetian. 2018. "Phytaspase-Mediated Precursor Processing and Maturation of the Wound Hormone Systemin." *New Phytologist* 218(3):1167–78.

Beneloujaephajri, Emna, Alex Costa, Floriane L'Haridon, Jean-Pierre Métraux, and Matteo Binda. 2013. "Production of Reactive Oxygen Species and Wound-Induced Resistance in Arabidopsis Thaliana against Botrytis Cinerea Are Preceded and Depend on a Burst of Calcium." *BMC Plant Biology* 13(1):160.

Benschop, Joris J., Shabaz Mohammed, Martina O'Flaherty, Albert J. R. Heck, Monique Slijper, and Frank L. H. Menke. 2007. "Quantitative Phosphoproteomics of Early Elicitor Signaling in Arabidopsis." *Molecular & Cellular Proteomics* 6(7):1198–1214.

Bergey, Daniel R. and Clarence A. Ryan. 1999. "Wound- and Systemin-Inducible Calmodulin Gene Expression in Tomato Leaves." *Plant Molecular Biology* 40(5):815–23.

Berriri, Souha, Ana Victoria Garcia, Nicolas Frei dit Frey, Wilfried Rozhon, Stéphanie Pateyron, Nathalie Leonhardt, Jean-Luc Montillet, Jeffrey Leung, Heribert Hirt, and Jean Colcombet. 2012. "Constitutively Active Mitogen-Activated Protein Kinase Versions Reveal Functions of Arabidopsis MPK4 in Pathogen Defense Signaling." *The Plant Cell* 24(10):4281–93.

Bishop, P. D., D. J. Makus, G. Pearce, and C. A. Ryan. 1981. "Proteinase Inhibitor-Inducing Factor Activity in Tomato Leaves Resides in Oligosaccharides Enzymically Released from Cell Walls." *Proceedings of the National Academy of Sciences* 78(6):3536–40.

Boller, Thomas and Georg Felix. 2009. "A Renaissance of Elicitors: Perception of Microbe-Associated Molecular Patterns and Danger Signals by Pattern-Recognition Receptors." *Annual Review of Plant Biology* 60(1):379–406.

Bosch, Marko, Sonja Berger, Andreas Schaller, and Annick Stintzi. 2014a. "Jasmonate-Dependent Induction of Polyphenol Oxidase Activity in Tomato Foliage Is Important for Defense against Spodoptera Exigua but Not against Manduca Sexta." *BMC Plant Biology* 14(1):257.

Bosch, Marko, Louwrance P. Wright, Jonathan Gershenzon, Claus Wasternack, Bettina Hause, Andreas Schaller, and Annick Stintzi. 2014b. "Jasmonic Acid and Its Precursor 12-Oxophytodienoic Acid Control Different Aspects of Constitutive and Induced Herbivore Defenses in Tomato." *Plant Physiology* 166(1):396–410.

Bowles, Dianna. 1990. "DEFENSE-RELATED PROTEINS IN HIGHER PLANTS." *Ann. Rev. Biochem.* 59:873–907.

Bowles, Dianna. 1998. "Signal Transduction in the Wound Response of Tomato Plants" edited by N. –H. Chua, A. M. Hetherington, R. Hooley, and R. F. Irvine. *Philosophical Transactions of the Royal Society of London. Series B: Biological Sciences* 353(1374):1495–1510.

Bradford, Marion M. 1976. "A Rapid and Sensitive Method for the Quantitation of Microgram Quantities of Protein Utilizing the Principle of Protein-Dye Binding." *Analytical Biochemistry* 72(1–2):248–54.

Bradshaw, Niels, Vladimir M. Levdikov, Christina M. Zimanyi, Rachelle Gaudet, Anthony J. Wilkinson, and Richard Losick. 2017. "A Widespread Family of Serine/Threonine Protein Phosphatases Shares a Common Regulatory Switch with Proteasomal Proteases." *ELife* 6:1–23.

Bubici, Giovanni, Anna Vittoria Carluccio, Livia Stavolone, and Fabrizio Cillo. 2017. "Prosystemin Overexpression Induces Transcriptional Modifications of Defense-Related and Receptor-like Kinase Genes and Reduces the Susceptibility to Cucumber Mosaic Virus and Its Satellite RNAs in Transgenic Tomato Plants" edited by T. P. Devarenne. *PLOS ONE* 12(2):e0171902.

Buist, Girbe, Anton Steen, Jan Kok, and Oscar P. Kuipers. 2008. "LysM, a Widely Distributed Protein Motif for Binding to (Peptido)Glycans." *Molecular Microbiology* 68(4):838–47.

Cao, Yangrong, Yan Liang, Kiwamu Tanaka, Cuong T. Nguyen, Robert P. Jedrzejczak, Andrzej Joachimiak, and Gary Stacey. 2014. "The Kinase LYK5 Is a Major Chitin Receptor in Arabidopsis and Forms a Chitin-Induced Complex with Related Kinase CERK1." *ELife* 3:1–19.

Cedzich, Anna, Franziska Huttenlocher, Benjamin M. Kuhn, Jens Pfannstiel, Leszek Gabier, Annick Stintzi, and Andreas Schaller. 2009. "The Protease-Associated Domain and C-Terminal Extension Are Required for Zymogen Processing, Sorting within the Secretory Pathway, and Activity of Tomato Subtilase 3 (SlSBT3)." *Journal of Biological Chemistry* 284(21):14068–78.

Chao, Wun S., Yong-Qiang Gu, Véronique Pautot, Elizabeth A. Bray, and Linda L. Walling. 1999. "Leucine Aminopeptidase RNAs, Proteins, and Activities Increase in Response to Water Deficit, Salinity, and the Wound Signals Systemin, Methyl Jasmonate, and Abscisic Acid." *Plant Physiology* 120(4):979–92.

Chassot, Céline, Antony Buchala, Henk Jan Schoonbeek, Jean Pierre Métraux, and Olivier Lamotte. 2008. "Wounding of Arabidopsis Leaves Causes a Powerful but Transient Protection against Botrytis Infection." *Plant Journal* 55(4):555–67.

Chen, Hui, C. G. Wilkerson, J. A. Kuchar, B. S. Phinney, and G. A. Howe. 2005. "Jasmonate-Inducible Plant Enzymes Degrade Essential Amino Acids in the Herbivore Midgut." *Proceedings of the National Academy of Sciences* 102(52):19237–42.

Chen, Yanmei and Wolfgang Hoehenwarter. 2015. "Changes in the Phosphoproteome and Metabolome Link Early Signaling Events to Rearrangement of Photosynthesis and Central Metabolism in Salinity and Oxidative Stress Response in Arabidopsis." *Plant Physiology* 169(December):pp.01486.2015.

Chen, Yanmei, Wolfgang Hoehenwarter, and Wolfram Weckwerth. 2010. "Comparative Analysis of Phytohormone-Responsive Phosphoproteins in Arabidopsis Thaliana Using Tio2-Phosphopeptide Enrichment and Mass Accuracy Precursor Alignment." *The Plant Journal* 63:1–17.

Cheng, Heung-Chin, Robert Z. Qi, Hemant Paudel, and Hong-Jian Zhu. 2011. "Regulation and Function of Protein Kinases and Phosphatases." *Enzyme Research* 2011:Article ID 794089.

Choi, Hyong Woo and Daniel F. Klessig. 2016. "DAMPs, MAMPs, and NAMPs in Plant Innate Immunity." *BMC Plant Biology* 16(1):232.

Choi, Jeongmin., K. Tanaka, Y. Cao, Y. Qi, J. Qiu, Y. Liang, S. Y. Lee, and G. Stacey. 2014. "Identification of a Plant Receptor for Extracellular ATP." *Science* 343(6168):290–94.

Chowdhury, Saikat Dutta and Ansuman Lahiri. 2017. "Plant Polypeptide Hormone Systemin Prefers Polyproline II Conformation in Solution." *ACS Omega* 2(10):6831–43.

Chung, S. H., C. Rosa, E. D. Scully, M. Peiffer, J. F. Tooker, K. Hoover, D. S. Luthe, and G. W. Felton. 2013. "Herbivore Exploits Orally Secreted Bacteria to Suppress Plant Defenses." *Proceedings of the National Academy of Sciences* 110(39):15728–33.

Constabel, C. Peter, Lynn Yip, and Clarence A. Ryan. 1998. "Prosystemin from Potato, Black Nightshade, and Bell Pepper: Primary Structure and Biological Activity of Predicted Systemin Polypeptides." *Plant Molecular Biology* 36(1):55–62.

Coppola, Mariangela, Giandomenico Corrado, Valentina Coppola, Pasquale Cascone, Rosanna Martinelli, Maria Cristina Digilio, Francesco Pennacchio, and Rosa Rao. 2015. "Prosystemin Overexpression in Tomato Enhances Resistance to Different Biotic Stresses by Activating Genes of Multiple Signaling Pathways." *Plant Molecular Biology Reporter* 33(5):1270–85.

Couto, Daniel, Roda Niebergall, Xiangxiu Liang, Christoph A. Bücherl, Jan Sklenar, Alberto P. Macho, Vardis Ntoukakis, Paul Derbyshire, Denise Altenbach, Dan Maclean, Silke Robatzek, Joachim Uhrig, Frank Menke, Jian Min Zhou, and Cyril Zipfel. 2016. "The Arabidopsis Protein Phosphatase PP2C38 Negatively Regulates the Central Immune Kinase BIK1." *PLoS Pathogens* 12(8):1–24.

Cox, Jürgen and Matthias Mann. 2008. "MaxQuant Enables High Peptide Identification Rates, Individualized p.p.b.-Range Mass Accuracies and Proteome-Wide Protein Quantification." *Nature Biotechnology* 26(12):1367–72.

Cox, Jürgen, Nadin Neuhauser, Annette Michalski, Richard A. Scheltema, Jesper V. Olsen, and Matthias Mann. 2011. "Andromeda: A Peptide Search Engine

Integrated into the MaxQuant Environment." *Journal of Proteome Research* 10(4):1794–1805.

Cui, Haitao, Kenichi Tsuda, and Jane E. Parker. 2015. "Effector-Triggered Immunity: From Pathogen Perception to Robust Defense." *Annual Review of Plant Biology* 66(1):487–511.

Czyzewicz, Nathan, Mari Wildhagen, Pietro Cattaneo, Yvonne Stahl, Karine Gustavo Pinto, Reidunn B. Aalen, Melinka A. Butenko, Rüdiger Simon, Christian S. Hardtke, and Ive De Smet. 2015. "Antagonistic Peptide Technology for Functional Dissection of CLE Peptides Revisited." *Journal of Experimental Botany* 66(17):5367–74.

Dalin, Peter and Christer Björkman. 2003. "Adult Beetle Grazing Induces Willow Trichome Defence against Subsequent Larval Feeding." *Oecologia* 134(1):112–18.

Degenhardt, David C., Sarah Refi-Hind, Johannes W. Stratmann, and David E. Lincoln. 2010. "Systemin and Jasmonic Acid Regulate Constitutive and Herbivore-Induced Systemic Volatile Emissions in Tomato, Solanum Lycopersicum." *Phytochemistry* 71(17–18):2024–37.

Denancé, Nicolas, Andrea Sánchez-Vallet, Deborah Goffner, and Antonio Molina. 2013. "Disease Resistance or Growth: The Role of Plant Hormones in Balancing Immune Responses and Fitness Costs." *Frontiers in Plant Science* 4(May):1–12.

Denoux, Carine, Roberta Galletti, Nicole Mammarella, Suresh Gopalan, Danièle Werck, Giulia De Lorenzo, Simone Ferrari, Frederick M. Ausubel, and Julia Dewdney. 2008. "Activation of Defense Response Pathways by OGs and Flg22 Elicitors in Arabidopsis Seedlings." *Molecular Plant* 1(3):423–45.

Deslandes, Laurent and Susana Rivas. 2012. "Catch Me If You Can: Bacterial Effectors and Plant Targets." *Trends in Plant Science* 17(11):644–55.

Devireddy, Amith R., Sara I. Zandalinas, Aurelio Gómez-Cadenas, Eduardo Blumwald, and Ron Mittler. 2018. "Coordinating the Overall Stomatal Response of Plants: Rapid Leaf-to-Leaf Communication during Light Stress." *Science Signaling* 11(518):1–9.

Dombrowski, James E. and Daniel R. Bergey. 2007. "Calcium Ions Enhance Systemin Activity and Play an Integral Role in the Wound Response." *Plant Science* 172(2):335–44.

Dreizler, Konrad. 2018. "Die Reifung Des Peptidhormons Systemin Durch Phytaspasen Und Ihre Bedeutung Für Die Wundsignaltransduktion in Der Tomate." University of Hohenheim.

Du, Minmin, Jiuhai Zhao, David T. W. Tzeng, Yuanyuan Liu, Lei Deng, Tianxia Yang, Qingzhe Zhai, Fangming Wu, Zhuo Huang, Ming Zhou, Qiaomei Wang, Qian Chen, Silin Zhong, Chang-Bao Li, and Chuanyou Li. 2017. "MYC2 Orchestrates a Hierarchical Transcriptional Cascade That Regulates Jasmonate-Mediated Plant Immunity in Tomato." *The Plant Cell* 29(8):1883–1906.

Duan, Guangyou, Dirk Walther, and Waltraud X. Schulze. 2013. "Reconstruction and Analysis of Nutrient-Induced Phosphorylation Networks in Arabidopsis Thaliana." *Frontiers in Plant Science* 4(December):1–15.

Dubiella, U., R. Lassig, T. Romeis, H. Seybold, C. P. Witte, E. Komander, G. Durian, and W. X. Schulze. 2013. "Calcium-Dependent Protein Kinase/NADPH Oxidase Activation Circuit Is Required for Rapid Defense Signal Propagation." *Proceedings of the National Academy of Sciences* 110(21):8744–49.

Ehrlich, Paul R. and Peter H. Raven. 1964. "Butterflies and Plants: A Study in Coevolution." *Evolution* 18(4):586.

Eichenseer, Herbert, M. Claravon Mathews, Jian L. Bi, J. Brad Murphy, and Gary W. Felton. 1999. "Salivary Glucose Oxidase: Multifunctional Roles ForHelicoverpa Zea?" *Archives of Insect Biochemistry and Physiology* 42(1):99–109.

Elmore, James Mitch and Gitta Coaker. 2011. "The Role of the Plasma Membrane H+-ATPase in Plant-Microbe Interactions." *Molecular Plant* 4(3):416–27.

Engelsberger, Wolfgang R. and Waltraud X. Schulze. 2012. "Nitrate and Ammonium Lead to Distinct Global Dynamic Phosphorylation Patterns When Resupplied to Nitrogen-Starved Arabidopsis Seedlings." *The Plant Journal* 69(6):978–95.

Engler, Carola, Ramona Gruetzner, Romy Kandzia, and Sylvestre Marillonnet. 2009. "Golden Gate Shuffling: A One-Pot DNA Shuffling Method Based on Type IIs Restriction Enzymes" edited by J. Peccoud. *PLoS ONE* 4(5):e5553.

Erb, Matthias, Stefan Meldau, and Gregg A. Howe. 2012. "Role of Phytohormones in Insect-Specific Plant Reactions." *Trends in Plant Science* 17(5):250–59.

Falhof, Janus, Jesper Torbøl Pedersen, Anja Thoe Fuglsang, and Michael Palmgren. 2016. "Plasma Membrane H+-ATPase Regulation in the Center of Plant Physiology." *Molecular Plant* 9(3):323–37.

Farmer, Edward E. and Clarence A. Ryan. 1992. "Octadecanoid Precursors of Jasmonic Acid Activate the Synthesis of Wound-Inducible Proteinase Inhibitors." *The Plant Cell* 4(2):129.

Felix, Georg and Thomas Boller. 1995. "Systemin Induces Rapid Ion Fluxes and Ethylene Biosynthesis in L Ycopersicon Peruvianum Cells." *The Plant Journal* 7(3):381–89.

Ferrari, Simone, Daniel V Savatin, Francesca Sicilia, Giovanna Gramegna, Felice Cervone, and Giulia De Lorenzo. 2013. "Oligogalacturonides: Plant Damage-Associated Molecular Patterns and Regulators of Growth and Development." *Frontiers in Plant Science* 4(March):Article 49.

Fowler, Jonathan H., Javier Narváez-Vásquez, Dale N. Aromdee, Véronique Pautot, Frances M. Holzer, and Linda L. Walling. 2009. "Leucine Aminopeptidase Regulates Defense and Wound Signaling in Tomato Downstream of Jasmonic Acid." *The Plant Cell* 21(4):1239–51.

Froehlich, J. E., A. Itoh, and G. A. Howe. 2001. "Tomato Allene Oxide Synthase and Fatty Acid Hydroperoxide Lyase, Two Cytochrome P450s Involved in Oxylipin Metabolism, Are Targeted to Different Membranes of Chloroplast Envelope." *Plant Physiology* 125(1):306–17.

Frost, C. J., M. C. Mescher, J. E. Carlson, and C. M. De Moraes. 2008. "Plant Defense Priming against Herbivores: Getting Ready for a Different Battle." *Plant Physiology* 146(3):818–24.

Fu, Zheng Qing and Xinnian Dong. 2013. "Systemic Acquired Resistance: Turning Local Infection into Global Defense." *Annual Review of Plant Biology* 64(1):839–63.

Fuglsang, Anja Thoe, Grete Tulinius, Na Cui, and Michael Gjedde Palmgren. 2006. "Protein Phosphatase 2A Scaffolding Subunit A Interacts with Plasma Membrane H+-ATPase C-Terminus in the Same Region as 14-3-3 Protein." *Physiologia Plantarum* 128(2):334–40.

Fürstenberg-Hägg, Joel, Mika Zagrobelny, and Søren Bak. 2013. "Plant Defense against Insect Herbivores." *International Journal of Molecular Sciences* 14(5):10242–97.

Genot, Baptiste, Julien Lang, Souha Berriri, Marie Garmier, Françoise Gilard, Stéphanie Pateyron, Katrien Haustraete, Dominique Van Der Streaten, Heribert Hirt, and Jean Colcombet. 2017. "Constitutively Active Arabidopsis MAP Kinase 3 Triggers Defense Responses Involving Salicylic Acid and SUMM2 Resistance Protein." *Plant Physiology* 174(2):1238–49.

Ghelis, Thanos. 2011. "Signal Processing by Protein Tyrosine Phosphorylation in Plants." *Plant Signaling and Behavior* 6(7):24–33.

Gonzales-Vigil, E., C. M. Bianchetti, G. N. Phillips, and G. A. Howe. 2011. "Adaptive Evolution of Threonine Deaminase in Plant Defense against Insect Herbivores." *Proceedings of the National Academy of Sciences* 108(14):5897–5902.

González-Teuber, Marcia and Martin Heil. 2009. "Nectar Chemistry Is Tailored for Both Attraction of Mutualists and Protection from Exploiters." *Plant Signaling & Behavior* 4(9):809–13.

Green, T. R. and C. A. Ryan. 1972. "Wound-Induced Proteinase Inhibitor in Plant Leaves: A Possible Defense Mechanism against Insects." *Science* 175(4023):776–77.

Gu, Yong-Qiang and Linda L. Walling. 2000. "Specificity of the Wound-Induced Leucine Aminopeptidase (LAP-A) of Tomato." *European Journal of Biochemistry* 267(4):1178–87.

Guan, Yihong, Qinfang Zhu, Delai Huang, Shuyi Zhao, Li Jan Lo, and Jinrong Peng. 2015. "An Equation to Estimate the Difference between Theoretically Predicted and SDS PAGE-Displayed Molecular Weights for an Acidic Peptide." *Scientific Reports* 5(1):13370.

Gust, Andrea A., Rory Pruitt, and Thorsten Nürnberger. 2017. "Sensing Danger: Key to Activating Plant Immunity." *Trends in Plant Science* 22(9):779–91.

Hampp, Christine, Andreas Richter, Sonia Osorio, Günther Zellnig, Alok Krishna Sinha, Alexandra Jammer, Alisdair R. Fernie, Bernhard Grimm, and Thomas Roitsch. 2012. "Establishment of a Photoautotrophic Cell Suspension Culture of Arabidopsis Thaliana for Photosynthetic, Metabolic, and Signaling Studies." *Molecular Plant* 5(2):524–27.

Hartmann, J., D. Linke, C. Bonniger, A. Tholey, and M. Sauter. 2015. "Conserved Phosphorylation Sites in the Activation Loop of the Arabidopsis Phytosulfokine Receptor PSKR1 Differentially Affect Kinase and Receptor Activity." *Biochemical Journal* 472(3):379–91.

Haruta, Miyoshi, William M. Gray, and Michael R. Sussman. 2015. "Regulation of the Plasma Membrane Proton Pump (H+-ATPase) by Phosphorylation." *Current Opinion in Plant Biology* 28:68–75.

Haruta, Miyoshi, G. Sabat, K. Stecker, B. B. Minkoff, and M. R. Sussman. 2014. "A Peptide Hormone and Its Receptor Protein Kinase Regulate Plant Cell Expansion Suppl." *Science* 343(6169):408–11.

Hayashi, Maki, Shin Ichiro Inoue, Koji Takahashi, and Toshinori Kinoshita. 2011. "Immunohistochemical Detection of Blue Light-Induced Phosphorylation of the Plasma Membrane H +-ATPase in Stomatal Guard Cells." *Plant and Cell Physiology* 52(7):1238–48.

He, Zhongqi and C. Wayne Honeycutt. 2005. "A Modified Molybdenum Blue Method for Orthophosphate Determination Suitable for Investigating Enzymatic Hydrolysis of Organic Phosphates." *Communications in Soil Science and Plant Analysis* 36:1373–83.

Hegenauer, Volker, Ursula Fürst, Bettina Kaiser, Matthew Smoker, Cyril Zipfel, Georg Felix, Mark Stahl, and Markus Albert. 2016. "Detection of the Plant Parasite Cuscuta Reflexa by a Tomato Cell Surface Receptor." *Science* 353(6298):478–81.

Heil, Martin, Sabine Greiner, Harald Meimberg, Ralf Krüger, Jean-louis Noyer, Günther Heubl, K. Eduard Linsenmair, and Wilhelm Boland. 2004. "Evolutionary Change from Induced to Constitutive Expression of an Indirect Plant Resistance." *Nature* 430(6996):205–8.

Heil, Martin and Jurriaan Ton. 2008. "Long-Distance Signalling in Plant Defence." *Trends in Plant Science* 13(6):264–72.

Heitz, T., D. R. Bergey, and C. A. Ryan. 1997. "A Gene Encoding a Chloroplast-Targeted Lipoxygenase in Tomato Leaves Is Transiently Induced by Wounding, Systemin, and Methyl Jasmonate." *Plant Physiology* 114(3):1085–93.

Hellens, Roger P., E. A. Edwards, Nicola R. Leyland, Samantha Bean, and Philip M. Mullineaux. 2000. "PGreen: A Versatile and Flexible Binary Ti Vector for Agrobacterium-Mediated Plant Transformation." *Plant Molecular Biology* 42(6):819–32.

Higgins, Rebecca, Thomas Lockwood, Susan Holley, Roopa Yalamanchili, and Johannes W. Stratmann. 2007. "Changes in Extracellular PH Are Neither Required nor Sufficient for Activation of Mitogen-Activated Protein Kinases (MAPKs) in Response to Systemin and Fusicoccin in Tomato." *Planta* 225(6):1535–46.

Ho, Hooi Ling. 2015. "Functional Roles of Plant Protein Kinases in Signal Transduction Pathways during Abiotic and Biotic Stress." *Journal of Biodiversity, Bioprospecting and Development* 2(2):1000147.

Hohmann, Ulrich, Kelvin Lau, and Michael Hothorn. 2017. "The Structural Basis of Ligand Perception and Signal Activation by Receptor Kinases." *Annual Review of Plant Biology* 68(1):109–37.

Holley, Susan R., Roopa D. Yalamanchili, Daniel S. Moura, Clarence a Ryan, and Johannes W. Stratmann. 2003. "Convergence of Signaling Pathways Induced by Systemin, Oligosaccharide Elicitors, and Ultraviolet-B Radiation at the Level of Mitogen-Activated Protein Kinases in Lycopersicon Peruvianum Suspension-Cultured Cells." *Plant Physiology* 132(4):1728–38.

Holton, Nicholas, Ana Caño-Delgado, Kate Harrison, Teresa Montoya, Joanne Chory, and Gerard J. Bishop. 2007. "Tomato BRASSINOSTEROID INSENSITIVE1 Is Required for Systemin-Induced Root Elongation in Solanum Pimpinellifolium but Is Not Essential for Wound Signaling." *The Plant Cell* 19(5):1709–17.

Holton, Nicholas, Vladimir Nekrasov, Pamela C. Ronald, and Cyril Zipfel. 2015. "The Phylogenetically-Related Pattern Recognition Receptors EFR and XA21 Recruit Similar Immune Signaling Components in Monocots and Dicots." *PLoS Pathogens* 11(1):1–22.

Hõrak, Hanna, Maija Sierla, Kadri Tõldsepp, Cun Wang, Yuh-Shuh Wang, Maris Nuhkat, Ervin Valk, Priit Pechter, Ebe Merilo, Jarkko Salojärvi, Kirk Overmyer, Mart Loog, Mikael Brosché, Julian I. Schroeder, Jaakko Kangasjärvi, and Hannes Kollist. 2016. "A Dominant Mutation in the HT1 Kinase Uncovers Roles of MAP Kinases and GHR1 in CO_2 -Induced Stomatal Closure." *The Plant Cell* 28(10):2493–2509.

Hou, Shuguo, Xin Wang, Donghua Chen, Xue Yang, Mei Wang, David Turrà, Antonio Di Pietro, and Wei Zhang. 2014. "The Secreted Peptide PIP1 Amplifies Immunity through Receptor-Like Kinase 7" edited by B. Tyler. *PLoS Pathogens* 10(9):e1004331.

Houston, Kelly, Matthew R. Tucker, Jamil Chowdhury, Neil Shirley, and Alan Little. 2016. "The Plant Cell Wall: A Complex and Dynamic Structure As Revealed by the Responses of Genes under Stress Conditions." *Frontiers in Plant Science* 7(August):1–18.

Howe, Gregg A. and Georg Jander. 2008. "Plant Immunity to Insect Herbivores." *Annual Review of Plant Biology* 59(1):41–66.

Howe, Gregg A., Jonathan Lightner, John Browse, and Clarence A. Ryan. 1996. "An Octadecanoid Pathway Mutant (JL5) of Tomato Is Compromised in Signaling for Defense against Insect Attack." *The Plant Cell* 8(11):2067.

Howe, Gregg A. and Andreas Schaller. 2008. "Direct Defenses in Plants and Their Induction by Wounding and Insect Herbivores." Pp. 7–29 in *Induced Plant Resistance to Herbivory*, edited by A. Schaller. Springer, Dordrecht.

Hua, Deping, Cun Wang, Junna He, Hui Liao, Ying Duan, Ziqiang Zhu, Yan Guo, Zhizhong Chen, and Zhizhong Gong. 2012. "A Plasma Membrane Receptor Kinase, GHR1, Mediates Abscisic Acid- and Hydrogen Peroxide-Regulated Stomatal Movement in Arabidopsis." *The Plant Cell* 24(6):2546–61.

Igarashi, Daisuke, Kenichi Tsuda, and Fumiaki Katagiri. 2012. "The Peptide Growth Factor, Phytosulfokine, Attenuates Pattern-Triggered Immunity." *Plant Journal* 71(2):194–204.

Invergo, Brandon M. and Pedro Beltrao. 2018. "Reconstructing Phosphorylation Signalling Networks from Quantitative Phosphoproteomic Data." *Essays In Biochemistry* 62(4):525–34.

Irving, Helen R., David M. Cahill, and Chris Gehring. 2018. "Moonlighting Proteins and Their Role in the Control of Signaling Microenvironments, as Exemplified by CGMP and Phytosulfokine Receptor 1 (PSKR1)." *Frontiers in Plant Science* 9:415.

Jones, Jonathan D. G. and Jeffery L. Dangl. 2006. "The Plant Immune System." *Nature* 444(7117):323–29.

Jones, Jonathan D. G., Russell E. Vance, and Jeffery L. Dangl. 2016. "Intracellular Innate Immune Surveillance Devices in Plants and Animals." *Science* 354(6316):1117.

Jorgensen, C. and Rune Linding. 2008. "Directional and Quantitative Phosphorylation Networks." *Briefings in Functional Genomics and Proteomics* 7(1):17–26.

Kadota, Yasuhiro, Thomas W. H. Liebrand, Yukihisa Goto, Jan Sklenar, Paul Derbyshire, Frank L. H. Menke, Miguel-Angel Torres, Antonio Molina, Cyril Zipfel, Gitta Coaker, and Ken Shirasu. 2019. "Quantitative Phosphoproteomic Analysis Reveals Common Regulatory Mechanisms between Effector- and PAMP-triggered Immunity in Plants." *New Phytologist* 221(4):2160–75.

Kadota, Yasuhiro, Ken Shirasu, and Cyril Zipfel. 2015. "Regulation of the NADPH Oxidase RBOHD during Plant Immunity." *Plant and Cell Physiology* 56(8):1472–80.

Kadota, Yasuhiro, Jan Sklenar, Paul Derbyshire, Lena Stransfeld, Shuta Asai, Vardis Ntoukakis, Jonathan DG Jones, Ken Shirasu, Frank Menke, Alexandra Jones, and Cyril Zipfel. 2014. "Direct Regulation of the NADPH Oxidase RBOHD by the PRR-Associated Kinase BIK1 during Plant Immunity." *Molecular Cell* 54(1):43–55.

Kandoth, Pramod Kaitheri, Stefanie Ranf, Suchita S. Pancholi, Sastry Jayanty, Michael D. Walla, Wayne Miller, Gregg a Howe, David E. Lincoln, and Johannes W. Stratmann. 2007. "Tomato MAPKs LeMPK1, LeMPK2, and LeMPK3 Function in the Systemin-Mediated Defense Response against Herbivorous Insects." *Proceedings of the National Academy of Sciences of the United States of America* 104(29):12205–10.

Kärkönen, Anna and Kazuyuki Kuchitsu. 2015. "Reactive Oxygen Species in Cell Wall Metabolism and Development in Plants." *Phytochemistry* 112(1):22–32.

Keinath, Nana F., Sylwia Kierszniowska, Justine Lorek, Gildas Bourdais, Sharon A. Kessler, Hiroko Shimosato-Asano, Ueli Grossniklaus, Waltraud X. Schulze, Silke Robatzek, and Ralph Panstruga. 2010. "PAMP (Pathogen-Associated Molecular Pattern)-Induced Changes in Plasma Membrane Compartmentalization Reveal Novel Components of Plant Immunity." *Journal of Biological Chemistry* 285(50):39140–49.

Kimura, Sachie, Cezary Waszczak, Kerri Hunter, and Michael Wrzaczek. 2017. "Bound by Fate: The Role of Reactive Oxygen Species in Receptor-Like Kinase Signaling." *The Plant Cell* 29(4):638–54.

Klein, T. M., T. Gradziel, M. E. Fromm, and J. C. Sanford. 1988. "Factors Influencing Gene Delivery into Zea Mays Cells by High–Velocity Microprojectiles." *Nature Biotechnology* 6(5):559–63.

Kohorn, Bruce D. 2016. "Cell Wall-Associated Kinases and Pectin Perception." *Journal of Experimental Botany* 67(2):489–94.

Koncz, Csaba and Jeff Schell. 1986. "The Promoter of TL-DNA Gene 5 Controls the Tissue-Specific Expression of Chimaeric Genes Carried by a Novel Type of Agrobacterium Binary Vector." *MGG Molecular & General Genetics* 204(3):383–96.

Krol, Elzbieta, Tobias Mentzel, Delphine Chinchilla, Thomas Boller, Georg Felix, Birgit Kemmerling, Sandra Postel, Michael Arents, Elena Jeworutzki, Khaled A. S. Al-Rasheid, Dirk Becker, and Rainer Hedrich. 2010. "Perception of the Arabidopsis Danger Signal Peptide 1 Involves the Pattern Recognition Receptor At PEPR1 and Its Close Homologue At PEPR2." *Journal of Biological Chemistry* 285(18):13471–79.

Kushalappa, Ajjamada C., Kalenahalli N. Yogendra, and Shailesh Karre. 2016. "Plant Innate Immune Response: Qualitative and Quantitative Resistance." *Critical Reviews in Plant Sciences* 35(1):38–55.

Kwezi, Lusisizwe, Oziniel Ruzvidzo, Janet I. Wheeler, Kershini Govender, Sylvana Iacuone, Philip E. Thompson, Chris Gehring, and Helen R. Irving. 2011. "The Phytosulfokine (PSK) Receptor Is Capable of Guanylate Cyclase Activity and Enabling Cyclic GMP-Dependent Signaling in Plants." *Journal of Biological Chemistry* 286(25):22580–88.

De La Fuente Van Bentem, Sergio, Elisabeth Roitinger, Dorothea Anrather, Edina Csaszar, and Heribert Hirt. 2006. "Phosphoproteomics as a Tool to Unravel Plant Regulatory Mechanisms." *Physiologia Plantarum* 126(1):110–19.

Ladwig, Friederike, Renate I. Dahlke, Nils Stührwohldt, Jens Hartmann, Klaus Harter, and Margret Sauter. 2015. "Phytosulfokine Regulates Growth in Arabidopsis through a Response Module at the Plasma Membrane That Includes CYCLIC NUCLEOTIDE-GATED CHANNEL17, H $^+$ -ATPase, and BAK1." *The Plant Cell* 27(6):1718–29.

Lanfermeijer, F. C., M. Staal, R. Malinowski, J. W. Stratmann, and J. T. M. Elzenga. 2007. "Micro-Electrode Flux Estimation Confirms That the Solanum Pimpinellifolium Cu3 Mutant Still Responds to Systemin." *PLANT PHYSIOLOGY* 146(1):129–39.

Larsen, Martin R., Tine E. Thingholm, Ole N. Jensen, Peter Roepstorff, and Thomas J. D. Jørgensen. 2005. "Highly Selective Enrichment of Phosphorylated Peptides from Peptide Mixtures Using Titanium Dioxide Microcolumns." *Molecular & Cellular Proteomics : MCP* 4(7):873–86.

León, José, Enrique Rojo, José J. Sánchez-Serrano, J. León, Enrique Rojo, and J. J. Sánchez-Serrano. 2001. "Wound Signalling in Plants." *Journal of Experimental Botany* 52(354):1–9.

Levin, Donald A. 1976. "The Chemical Defenses of Plants to Pathogens and Herbivores." *Annual Review of Ecology and Systematics* 7(1):121–59.

Li, Chuanyou, Guanghui Liu, Changcheng Xu, Gyu In Lee, Petra Bauer, Hong-Qing Ling, Martin W. Ganal, and Gregg A. Howe. 2003. "The Tomato Suppressor of Prosystemin-Mediated Responses2 Gene Encodes a Fatty Acid Desaturase Required for the Biosynthesis of Jasmonic Acid and the Production of a Systemic Wound Signal for Defense Gene Expression." *The Plant Cell* 15(July):1646–1661.

Li, Chuanyou, Anthony L. Schilmiller, Guanghui Liu, Gyu In Lee, Sastry Jayanty, Carolyn Sageman, Julia Vrebalov, James J. Giovannoni, Kaori Yagi, Yuichi Kobayashi, and Gregg A. Howe. 2005. "Role of Beta-Oxidation in Jasmonate Biosynthesis and Systemic Wound Signaling in Tomato." *The Plant Cell* 17(3):971–86.

Li, L., C. Li, G. I. Lee, and G. A. Howe. 2002. "Distinct Roles for Jasmonate Synthesis and Action in the Systemic Wound Response of Tomato." *Proceedings of the National Academy of Sciences* 99(9):6416–21.

Li, L., Chuanyou Li, and G. A. Howe. 2001. "Genetic Analysis of Wound Signaling in Tomato. Evidence for a Dual Role of Jasmonic Acid in Defense and Female Fertility." *PLANT PHYSIOLOGY* 127(4):1414–17.

Li, Xiao-Shui, Bi-Feng Yuan, and Yu-Qi Feng. 2016. "Recent Advances in Phosphopeptide Enrichment: Strategies and Techniques." *Trends in Analytical Chemistry* 78:70–83.

Lim, Chae Woo, Seung Hwan Yang, Ki Hun Shin, Sung Chul Lee, and Sang Hyon Kim. 2014. "The AtLRK10L1.2, Arabidopsis Ortholog of Wheat LRK10, Is

Involved in ABA-Mediated Signaling and Drought Resistance." *Plant Cell Reports* 34(3):447–55.

Lin, Li-Ling, Chia-Lang Hsu, Chia-Wei Hu, Shiao-Yun Ko, Hsu-Liang Hsieh, Hsuan-Cheng Huang, and Hsueh-Fen Juan. 2015. "Integrating Phosphoproteomics and Bioinformatics to Study Brassinosteroid-Regulated Phosphorylation Dynamics in Arabidopsis." *BMC Genomics* 16(1):533.

Loivamäki, Maaria, Nils Stührwohldt, Rosalia Deeken, Bianka Steffens, Thomas Roitsch, Rainer Hedrich, and Margret Sauter. 2010. "A Role for PSK Signaling in Wounding and Microbial Interactions in Arabidopsis." *Physiologia Plantarum* 139(4):348–57.

LoPresti, Eric F. 2016. "Chemicals on Plant Surfaces as a Heretofore Unrecognized, but Ecologically Informative, Class for Investigations into Plant Defence." *Biological Reviews* 91(4):1102–17.

Lorbiecke, René and Margret Sauter. 2002. "Comparative Analysis of PSK Peptide Growth Factor Precursor Homologs." *Plant Science* 163(2):321–32.

Lortzing, Tobias and Anke Steppuhn. 2016. "Jasmonate Signalling in Plants Shapes Plant-Insect Interaction Ecology." *Current Opinion in Insect Science* 14:32–39.

Luan, Sheng. 2003. "PROTEIN PHOSPHATASES IN PLANTS." *Annual Review of Plant Biology* 54(1):63–92.

Macho, Alberto P. and Cyril Zipfel. 2014. "Plant PRRs and the Activation of Innate Immune Signaling." *Molecular Cell* 54(2):263–72.

Manosalva, Patricia, Murli Manohar, Stephan H. von Reuss, Shiyan Chen, Aline Koch, Fatma Kaplan, Andrea Choe, Robert J. Micikas, Xiaohong Wang, Karl-Heinz Kogel, Paul W. Sternberg, Valerie M. Williamson, Frank C. Schroeder, and Daniel F. Klessig. 2015. "Conserved Nematode Signalling Molecules Elicit Plant Defenses and Pathogen Resistance." *Nature Communications* 6(1):7795.

Martin, Maud, Richard Kettmann, and Franck Dequiedt. 2010. "Recent Insights into Protein Phosphatase 2A Structure and Regulation : The Reasons Why PP2A Is No Longer Considered as a Lazy Passive Housekeeping Enzyme." *Biotechnology, Agronomy, Society and Environment* 14(1):243–52.

Matsubayashi, Yoshikatsu. 2014. "Posttranslationally Modified Small-Peptide Signals in Plants." *Annual Review of Plant Biology* 65(1):385–413.

Matsubayashi, Yoshikatsu, Mari Ogawa, Hitomi Kihara, Masaaki Niwa, and Youji Sakagami. 2006. "Disruption and Overexpression of Arabidopsis Phytosulfokine Receptor Gene Affects Cellular Longevity and Potential for Growth." *Plant Physiology* 142(1):45–53.

Matsubayashi, Yoshikatsu, Mari Ogawa, Akiko Morita, and Youji Sakagami. 2002. "An LRR Receptor Kinase Involved in Perception of a Peptide Plant Hormone, Phytosulfokine." *Science (New York, N.Y.)* 296(5572):1470–72.

Matsubayashi, Yoshikatsu and Youji Sakagami. 1996. "Phytosulfokine, Sulfated
Peptides That Induce the Proliferation of Single Mesophyll Cells of Asparagus
Officinalis L." *Proceedings of the National Academy of Sciences* 93(15):7623–27.

Mattei, Benedetta, Francesco Spinelli, Daniela Pontiggia, and Giulia De Lorenzo.
2016. "Comprehensive Analysis of the Membrane Phosphoproteome Regulated
by Oligogalacturonides in Arabidopsis Thaliana." *Frontiers in Plant Science*
7(August):Article 1107.

Mattiacci, Letizia, Marcel Dicke, and Maarten A. Posthumust. 1995. "Beta-
Glucosidase: An Elicitor of Herbivore-Induced Plant Odor That Attracts Host-
Searching Parasitic Wasps." *Proceedings of the National Academy of Sciences of
the United States of America* 92(March):2036–40.

Mauch-Mani, Brigitte, Ivan Baccelli, Estrella Luna, and Victor Flors. 2017. "Defense
Priming: An Adaptive Part of Induced Resistance." *Annual Review of Plant
Biology* 68(1):485–512.

McGurl, Barry, Martha Orozcocardenas, Gregory Pearce, and Clarence A. Ryant.
1994. "Overexpression of the Prosystemin Gene in Transgenic Tomato Plants
Generates a Systemic Signal That Constitutively Induces Proteinase Inhibitor
Synthesis." *Plant Biology* 91(21):9799–9802.

McGurl, Barry, Gregory Pearce, Martha Orozco-cardenas, and Clarence A. Ryan.
1992. "Structure, Expression, and Antisense Inhibition of the Systemin Precursor
Gene." *Science* 255(5051):1570–74.

McGurl, Barry and Clarence A. Ryan. 1992. "The Organization of the Prosystemin
Gene." *Plant Molecular Biology* 20(3):405–9.

Meindl, T., T. Boller, and G. Felix. 1998. "The Plant Wound Hormone Systemin
Binds with the N-Terminal Part to Its Receptor but Needs the C-Terminal Part to
Activate It." *The Plant Cell* 10(9):1561–70.

Mello, Marcia O. and Marcio C. Silva-Filho. 2002. "Plant-Insect Interactions: An
Evolutionary Arms Race between Two Distinct Defense Mechanisms." *Brazilian
Journal of Plant Physiology* 14(2):71–81.

Mersmann, Sophia, Gildas Bourdais, Steffen Rietz, and Silke Robatzek. 2010.
"Ethylene Signaling Regulates Accumulation of the FLS2 Receptor and Is
Required for the Oxidative Burst Contributing to Plant Immunity." *Plant
Physiology* 154(1):391–400.

Miersch, O. and C. Wasternack. 2000. "Octadecanoid and Jasmonate Signaling in
Tomato (Lycopersicon Esculentum Mill.) Leaves: Endogenous Jasmonates Do
Not Induce Jasmonate Biosynthesis." *Biological Chemistry* 381(8):715–22.

Mitchell, Carolyn, Rex M. Brennan, Julie Graham, and Alison J. Karley. 2016. "Plant
Defense against Herbivorous Pests: Exploiting Resistance and Tolerance Traits
for Sustainable Crop Protection." *Frontiers in Plant Science* 7(July):1–8.

Mithoe, Sharon C., Christina Ludwig, Michiel JC Pel, Mara Cucinotta, Alberto
Casartelli, Malick Mbengue, Jan Sklenar, Paul Derbyshire, Silke Robatzek, Corné

134

MJ Pieterse, Ruedi Aebersold, and Frank LH Menke. 2016. "Attenuation of Pattern Recognition Receptor Signaling Is Mediated by a MAP Kinase Kinase Kinase." *EMBO Reports* 17(3):441–54.

Mithöfer, Axel. and Wilhelm. Boland. 2008. "Recognition of Herbivory-Associated Molecular Patterns." *Plant Physiology* 146(3):825–31.

Mithöfer, Axel and Wilhelm Boland. 2012. "Plant Defense Against Herbivores: Chemical Aspects." *Annual Review of Plant Biology* 63(1):431–50.

Mittler, Ron and Eduardo Blumwald. 2015. "The Roles of ROS and ABA in Systemic Acquired Acclimation." *The Plant Cell Online* 27(1):64–70.

Mosher, Stephen and Birgit Kemmerling. 2013. "PSKR1 and PSY1R-Mediated Regulation of Plant Defense Responses." *Plant Signaling and Behavior* 8(5):e24119.

Mosher, Stephen, Heike Seybold, Patricia Rodriguez, Mark Stahl, Kelli A. Davies, Sajeewani Dayaratne, Santiago A. Morillo, Michael Wierzba, Bruno Favery, Harald Keller, Frans E. Tax, and Birgit Kemmerling. 2013. "The Tyrosine-Sulfated Peptide Receptors PSKR1 and PSY1R Modify the Immunity of Arabidopsis to Biotrophic and Necrotrophic Pathogens in an Antagonistic Manner." *Plant Journal* 73(3):469–82.

Motose, H., K. Iwamoto, S. Endo, T. Demura, Y. Sakagami, Y. Matsubayashi, K. L. Moore, and H. Fukuda. 2009. "Involvement of Phytosulfokine in the Attenuation of Stress Response during the Transdifferentiation of Zinnia Mesophyll Cells into Tracheary Elements." *Plant Physiology* 150(1):437–47.

Mumby, Marc and Deirdre Brekken. 2005. "Phosphoproteomics: New Insights into Cellular Signaling." *Genome Biology* 6(9):Article 230.

Murphy, James M., Qingwei Zhang, Samuel N. Young, Michael L. Reese, Fiona P. Bailey, Patrick A. Eyers, Daniela Ungureanu, Henrik Hammaren, Olli Silvennoinen, Leila N. Varghese, Kelan Chen, Anne Tripaydonis, Natalia Jura, Koichi Fukuda, Jun Qin, Zachary Nimchuk, Mary Beth Mudgett, Sabine Elowe, Christine L. Gee, Ling Liu, Roger J. Daly, Gerard Manning, Jeffrey J. Babon, and Isabelle S. Lucet. 2014. "A Robust Methodology to Subclassify Pseudokinases Based on Their Nucleotide-Binding Properties." *Biochemical Journal* 457(2):323–34.

Muthamilarasan, Mehanathan and Manoj Prasad. 2013. "Plant Innate Immunity: An Updated Insight into Defense Mechanism." *Journal of Biosciences* 38(2):433–49.

Nakagami, Hirofumi, Naoyuki Sugiyama, Yasushi Ishihama, and Ken Shirasu. 2012. "Shotguns in the Front Line: Phosphoproteomics in Plants." *Plant and Cell Physiology* 53(1):118–24.

Narvaez-Vasquez, Javier, Jorge Florin-Christensen, and Clarence A. Ryan. 1999. "Positional Specificity of a Phospholipase A Activity Induced by Wounding, Systemin, and Oligosaccharide Elicitors in Tomato Leaves." *The Plant Cell* 11(11):2249.

Narváez-Vásquez, Javier and Clarence A. Ryan. 2004. "The Cellular Localization of Prosystemin: A Functional Role for Phloem Parenchyma in Systemic Wound Signaling." *Planta* 218(3):360–69.

Nie, Wen-Feng, Meng-Meng Wang, XIiao-Jian Xia, Yan-Hong Zhou, Kai Shi, Zhixiang Chen, and JIing Quan Yu. 2013. "Silencing of Tomato RBOH1 and MPK2 Abolishes Brassinosteroid-Induced H2O2 Generation and Stress Tolerance." *Plant, Cell & Environment* 36(4):789–803.

Niittylä, Totte, Anja T. Fuglsang, Michael G. Palmgren, Wolf B. Frommer, and Waltraud X. Schulze. 2007. "Temporal Analysis of Sucrose-Induced Phosphorylation Changes in Plasma Membrane Proteins of *Arabidopsis*." *Molecular & Cellular Proteomics* 6(10):1711–26.

Nühse, Thomas S., Andrew R. Bottrill, Alexandra M. E. Jones, and Scott C. Peck. 2007. "Quantitative Phosphoproteomic Analysis of Plasma Membrane Proteins Reveals Regulatory Mechanisms of Plant Innate Immune Responses." *The Plant Journal* 51(5):931–40.

Nühse, Thomas S., Scott C. Peck, Heribert Hirt, and Thomas Boller. 2000. "Microbial Elicitors Induce Activation and Dual Phosphorylation of the Arabidopsis Thaliana MAPK 6." *Journal of Biological Chemistry* 275(11):7521–26.

Nühse, Thomas S., Allan Stensballe, Ole N. Jensen, and Scott C. Pecka. 2004. "Phosphoproteomics of the Arabidopsis Plasma Membrane and a New Phosphorylation Site Database." *The Plant Cell* Vol. 16(September):2394–2405.

O'Donnell, P. J., C. Calvert, R. Atzorn, C. Wasternack, H. M. O. Leyser, and D. J. Bowles. 1996. "Ethylene as a Signal Mediating the Wound Response of Tomato Plants." *Science* 274(5294):1914–17.

Ogasawara, Yoko, Hidetaka Kaya, Goro Hiraoka, Fumiaki Yumoto, Sachie Kimura, Yasuhiro Kadota, Haruka Hishinuma, Eriko Senzaki, Satoshi Yamagoe, Koji Nagata, Masayuki Nara, Kazuo Suzuki, Masaru Tanokura, and Kazuyuki Kuchitsu. 2008. "Synergistic Activation of the Arabidopsis NADPH Oxidase AtrbohD by Ca 2+ and Phosphorylation." *Journal of Biological Chemistry* 283(14):8885–92.

Orozco-Cardenas, Martha L., B. McGurl, and C. A. Ryan. 2006. "Expression of an Antisense Prosystemin Gene in Tomato Plants Reduces Resistance toward Manduca Sexta Larvae." *Proceedings of the National Academy of Sciences* 90(17):8273–76.

Orozco-Cardenas, Martha L., Javier Narvaez-Vasquez, and Clarence A. Ryan. 2001. "Hydrogen Peroxide Acts as a Second Messenger for the Induction of Defense Genes in Tomato Plants in Response to Wounding, Systemin, and Methyl Jasmonate." *The Plant Cell* 13(1):179.

Orozco-Cardenas, Martha L. and C. A. Ryan. 1999. "Hydrogen Peroxide Is Generated Systemically in Plant Leaves by Wounding and Systemin via the Octadecanoid Pathway." *Proceedings of the National Academy of Sciences* 96(11):6553–57.

Orsburn, Benjamin C., Luke H. Stockwin, and Dianne L. Newton. 2011. "Challenges in Plasma Membrane Phosphoproteomics." *Expert Review of Proteomics* 8(4):483–94.

Park, Chang-Jin, Ying Peng, Xuewei Chen, Christopher Dardick, DeLing Ruan, Rebecca Bart, Patrick E. Canlas, and Pamela C. Ronald. 2008. "Rice XB15, a Protein Phosphatase 2C, Negatively Regulates Cell Death and XA21-Mediated Innate Immunity" edited by J. L. Dangl. *PLoS Biology* 6(9):e231.

Pastor, Victoria, Paloma Sánchez-Bel, Jordi Gamir, María J. Pozo, and Víctor Flors. 2018. "Accurate and Easy Method for Systemin Quantification and Examining Metabolic Changes under Different Endogenous Levels." *Plant Methods* 14(1):1–14.

Pearce, Gregory. 2011. "Systemin, Hydroxyproline-Rich Systemin and the Induction of Protease Inhibitors." *Current Protein & Peptide Science* 12(5):399–408.

Pearce, Gregory, S. Johnson, and C. A. Ryan. 1993. "Structure-Activity of Deleted and Substituted Systemin, an 18-Amino Acid Polypeptide Inducer of Plant Defensive Genes." *Journal of Biological Chemistry* 268(1):212–16.

Pearce, Gregory, D. Strdom, S. Johnson, and C. A. Ryan. 1991. "A Polypeptide from Tomato Leaves Induces Wound-Inducible Proteinase Inhibitor Proteins." *Science* 253(5022):895–97.

Peck, Scott C. 2003. "Early Phosphorylation Events in Biotic Stress." *Current Opinion in Plant Biology* 6(4):334–38.

Pedley, Kerry F. and Gregory B. Martin. 2004. "Identification of MAPKs and Their Possible MAPK Kinase Activators Involved in the Pto-Mediated Defense Response of Tomato." *Journal of Biological Chemistry* 279(47):49229–35.

Pedley, Kerry F. and Gregory B. Martin. 2005. "Role of Mitogen-Activated Protein Kinases in Plant Immunity." *Current Opinion in Plant Biology* 8(5):541–47.

Peña-Cortés, H., Joachim Fisahn, and Lothar Willmitzer. 1995. "Signals Involved in Wound-Induced Proteinase Inhibitor II Gene Expression in Tomato and Potato Plants." *Proceedings of the National Academy of Sciences of the United States of America* 92(10):4106–13.

Petutschnig, Elena K., Alexandra M. E. Jones, Liliya Serazetdinova, Ulrike Lipka, and Volker Lipka. 2010. "The Lysin Motif Receptor-like Kinase (LysM-RLK) CERK1 Is a Major Chitin-Binding Protein in Arabidopsis Thaliana and Subject to Chitin-Induced Phosphorylation." *Journal of Biological Chemistry* 285(37):28902–11.

Pieterse, Corné M. J., Dieuwertje Van der Does, Christos Zamioudis, Antonio Leon-Reyes, and Saskia C. M. Van Wees. 2012. "Hormonal Modulation of Plant Immunity." *Annual Review of Cell and Developmental Biology* 28(1):489–521.

Rampitsch, Christof and Natalia V. Bykova. 2012. "The Beginnings of Crop Phosphoproteomics: Exploring Early Warning Systems of Stress." *Frontiers in Plant Science* 3(July):1–15.

Ranf, Stefanie, Lennart Eschen-Lippold, Pascal Pecher, Justin Lee, and Dierk Scheel. 2011. "Interplay between Calcium Signalling and Early Signalling Elements during Defence Responses to Microbe- or Damage-Associated Molecular Patterns." *The Plant Journal* 68(1):100–113.

Rappsilber, Juri, Matthias Mann, and Yasushi Ishihama. 2007. "Protocol for Micro-Purification, Enrichment, Pre-Fractionation and Storage of Peptides for Proteomics Using StageTips." *Nature Protocols* 2(8):1896–1906.

Rayapuram, Naganand, Ludovic Bonhomme, Jean Bigeard, Kahina Haddadou, Cédric Przybylski, Heribert Hirt, and Delphine Pflieger. 2014. "Identification of Novel PAMP-Triggered Phosphorylation and Dephosphorylation Events in Arabidopsis Thaliana by Quantitative Phosphoproteomic Analysis." *Journal of Proteome Research* 13(4):2137–51.

Ryan, Clarence A. 2000. "The Systemin Signaling Pathway: Differential Activation of Plant Defensive Genes." *Biochimica et Biophysica Acta - Protein Structure and Molecular Enzymology* 1477(1–2):112–21.

Sagi, Moshe, Olga Davydov, Saltanat Orazova, Zhazira Yesbergenova, Ron Ophir, Johannes W. Stratmann, and Robert Fluhr. 2004. "Plant Respiratory Burst Oxidase Homologs Impinge on Wound Responsiveness and Development in Lycopersicon Esculentum." *The Plant Cell* 16(3):616–28.

Sakamoto, Tetsu, Michihito Deguchi, Otávio J. B. Brustolini, Anésia A. Santos, Fabyano F. Silva, and Elizabeth P. B. Fontes. 2012. "The Tomato RLK Superfamily: Phylogeny and Functional Predictions about the Role of the LRRII-RLK Subfamily in Antiviral Defense." *BMC Plant Biology* 12(229):1–18.

Salovska, Barbora, Ales Tichy, Martina Rezacova, Jirina Vavrova, and Eva Novotna. 2012. "Enrichment Strategies for Phosphoproteomics: State-of-the-Art." *Reviews in Analytical Chemistry* 31(1):29–41.

Sauter, Margret. 2015. "Phytosulfokine Peptide Signalling." *Journal of Experimental Botany* 66(17):5161–69.

Savatin, Daniel V, Giovanna Gramegna, Vanessa Modesti, and Felice Cervone. 2014. "Wounding in the Plant Tissue: The Defense of a Dangerous Passage." *Frontiers in Plant Science* 5(September):470.

Schaller, Andreas. 1998. "Action of Proteolysis-Resistant Systemin Analogues in Wound Signalling." *Phytochemistry* 47(4):605–12.

Schaller, Andreas. 1999. "Oligopeptide Signalling and the Action of Systemin." *Plant Molecular Biology* 40(5):763–69.

Schaller, Andreas and C. Oecking. 1999. "Modulation of Plasma Membrane H+-ATPase Activity Differentially Activates Wound and Pathogen Defense Responses in Tomato Plants." *The Plant Cell* 11(2):263–72.

Schaller, Andreas and Clarence A. Ryan. 1996. "Systemin-a Polypeptide Defense Signal in Plants." *BioEssays : News and Reviews in Molecular, Cellular and Developmental Biology* 18(1):27–33.

Schaller, Andreas and Annick Stintzi. 2008. "Jasmonate Biosynthesis and Signaling for Induced Plant Defense against Herbivory." Pp. 349–66 in *Induced Plant Resistance to Herbivory*. Dordrecht: Springer Netherlands.

Scheer, Justin M. and Ca Ryan. 1999. "A 160-KD Systemin Receptor on the Surface of Lycopersicon Peruvianum Suspension-Cultured Cells." *The Plant Cell* 11(8):1525–36.

Scheer, Justin M. and Clarence A. Ryan. 2002. "The Systemin Receptor SR160 from Lycopersicon Peruvianum Is a Member of the LRR Receptor Kinase Family." *Proceedings of the National Academy of Sciences of the United States of America* 99(14):9585–90.

Schenk, S. T., C. Hernandez-Reyes, B. Samans, E. Stein, C. Neumann, M. Schikora, M. Reichelt, A. Mithofer, A. Becker, K. H. Kogel, and A. Schikora. 2014. "N-Acyl-Homoserine Lactone Primes Plants for Cell Wall Reinforcement and Induces Resistance to Bacterial Pathogens via the Salicylic Acid/Oxylipin Pathway." *The Plant Cell* 26(6):2708–23.

Schilmiller, Anthony. L., I. Schauvinhold, M. Larson, R. Xu, A. L. Charbonneau, A. Schmidt, C. Wilkerson, R. L. Last, and E. Pichersky. 2009. "Monoterpenes in the Glandular Trichomes of Tomato Are Synthesized from a Neryl Diphosphate Precursor Rather than Geranyl Diphosphate." *Proceedings of the National Academy of Sciences* 106(26):10865–70.

Schilmiller, Anthony L. and Gregg A. Howe. 2005. "Systemic Signaling in the Wound Response." *Current Opinion in Plant Biology* 8(4):369–77.

Schmelz, Eric. A., M. J. Carroll, S. LeClere, S. M. Phipps, J. Meredith, P. S. Chourey, H. T. Alborn, and P. E. A. Teal. 2006. "Fragments of ATP Synthase Mediate Plant Perception of Insect Attack." *Proceedings of the National Academy of Sciences* 103(23):8894–99.

Schmelz, Eric A. 2015. "Impacts of Insect Oral Secretions on Defoliation-Induced Plant Defense." *Current Opinion in Insect Science* 9:7–15.

Schneider, Caroline A., Wayne S. Rasband, and Kevin W. Eliceiri. 2012. "NIH Image to ImageJ: 25 Years of Image Analysis." *Nature Methods* 9(7):671–75.

Schulze, Birgit, Tobias Mentzel, Anna K. Jehle, Katharina Mueller, Seraina Beeler, Thomas Boller, Georg Felix, and Delphine Chinchilla. 2010. "Rapid Heteromerization and Phosphorylation of Ligand-Activated Plant Transmembrane Receptors and Their Associated Kinase BAK1." *Journal of Biological Chemistry* 285(13):9444–51.

Shi, Yigong. 2009. "Serine/Threonine Phosphatases: Mechanism through Structure." *Cell* 139(3):468–84.

Shin, Dong M., Tae-Wook Chung, Fadlo R. Khuri, Jianxin Xie, Taro Hitosugi, Changliang Shan, Mike Aguiar, Shannon Elf, Ting-Lei Gu, Jun Fan, Hanna J. Khoury, Hee-Bum Kang, Titus J. Boggon, Martha Arellano, and Scott Lonning.

2014. "Tyr-94 Phosphorylation Inhibits Pyruvate Dehydrogenase Phosphatase 1 and Promotes Tumor Growth." *Journal of Biological Chemistry* 289(31):21413–22.

Shin, Ki Hun, Seung Hwan Yang, Jun Yong Lee, Che Woo Lim, Sung Chul Lee, John W. S. Brown, and Sang Hyon Kim. 2015. "Alternative Splicing of Mini-Exons in the Arabidopsis Leaf Rust Receptor-like Kinase LRK10 Genes Affects Subcellular Localisation." *Plant Cell Reports* 34(3):495–505.

Shinya, Tomonori, Shigetaka Yasuda, Kiwamu Hyodo, Rena Tani, Yuko Hojo, Yuka Fujiwara, Kei Hiruma, Takuma Ishizaki, Yasunari Fujita, Yusuke Saijo, and Ivan Galis. 2018. "Integration of Danger Peptide Signals with Herbivore-Associated Molecular Pattern Signaling Amplifies Anti-Herbivore Defense Responses in Rice." *The Plant Journal* 94(4):626–37.

Sierla, Maija, Hanna Hõrak, Kirk Overmyer, Cezary Waszczak, Dmitry Yarmolinsky, Tobias Maierhofer, Julia P. Vainonen, Konstantin Denessiouk, Jarkko Salojärvi, Kristiina Laanemets, Kadri Tõldsepp, Triin Vahisalu, Adrien Gauthier, Tuomas Puukko, Lars Paulin, Petri Auvinen, Dietmar Geiger, Rainer Hedrich, Hannes Kollist, and Jaakko Kangasjärvi. 2018. "The Receptor-like Pseudokinase GHR1 Is Required for Stomatal Closure." *The Plant Cell* 30(November):2813–2837.

Sivasankar, Sobhana, Bay Sheldrick, and Steven J. Rothstein. 2000. "Expression of Allene Oxide Synthase Determines Defense Gene Activation in Tomato." *Plant Physiology* 122(4):1335–42.

Song, X. F., P. Guo, T. T. Xu, C. M. Liu, and S. C. Ren. 2013. "Antagonistic Peptide Technology for Functional Dissection of CLV3/ESR Genes in Arabidopsis." *PLANT PHYSIOLOGY* 161(3):1076–85.

Song, Xiu Fen, Shi Chao Ren, and Chun Ming Liu. 2017. "Peptide Hormones." Pp. 361–404 in *Hormone Metabolism and Signaling in Plants*. Jiayang Li, Chuanyou Li and Steven Smith.

Spartz, A. K., H. Ren, M. Y. Park, K. N. Grandt, S. H. Lee, A. S. Murphy, M. R. Sussman, P. J. Overvoorde, and W. M. Gray. 2014. "SAUR Inhibition of PP2C-D Phosphatases Activates Plasma Membrane H+-ATPases to Promote Cell Expansion in Arabidopsis." *The Plant Cell* 26(5):2129–42.

Spoel, Steven H. and Xinnian Dong. 2008. "Making Sense of Hormone Crosstalk during Plant Immune Responses." *Cell Host and Microbe* 3(6):348–51.

Srivastava, Renu, Jian-Xiang Liu, and Stephen H. Howell. 2008. "Proteolytic Processing of a Precursor Protein for a Growth-Promoting Peptide by a Subtilisin Serine Protease in Arabidopsis." *The Plant Journal* 56(2):219–27.

Stanković, Bratislav and Eric Davies. 1997. "Intercellular Communication in Plants: Electrical Stimulation of Proteinase Inhibitor Gene Expression in Tomato." *Planta* 202(4):402–6.

Staswick, Paul E. and Iskender Tiryaki. 2004. "The Oxylipin Signal Jasmonic Acid Is Activated by an Enzyme That Conjugates It to Isoleucine in Arabidopsis." *The Plant Cell* 16(8):2117–27.

Stegmann, Martin, Jacqueline Monaghan, Elwira Smakowska-Luzan, Hanna Rovenich, Anita Lehner, Nicholas Holton, Youssef Belkhadir, and Cyril Zipfel. 2017. "The Receptor Kinase FER Is a RALF-Regulated Scaffold Controlling Plant Immune Signaling." *Science* 355(6322):287–89.

Steinite, Ineta, Agnese Gailite, and Gederts Ievinsh. 2004. "Reactive Oxygen and Ethylene Are Involved in the Regulation of Regurgitant-Induced Responses in Bean Plants." *Journal of Plant Physiology* 161(2):191–96.

Strassner, Jochen, Yoann Huet, and Andreas Schaller. 2002a. "Cloning of Tomato Proteases by Direct Selection in Yeast for Enzymes That Cleave the Polypeptide Wound Signal Systemin. In: Induced Resistance in Plants against Insects and Diseases." *Induced Resistance in Plants against Insects and Diseases* 25(6):159–63.

Strassner, Jochen, Florian Schaller, Ursula B. Frick, Gregg A. Howe, Elmar W. Weiler, Nikolaus Amrhein, Peter Macheroux, and Andreas Schaller. 2002b. "Characterization and CDNA-Microarray Expression Analysis of 12-Oxophytodienoate Reductases Reveals Differential Roles for Octadecanoid Biosynthesis in the Local versus the Systemic Wound Response." *The Plant Journal* 32(4):585–601.

Stührwohldt, Nils, Renate I. Dahlke, Bianka Steffens, Amanda Johnson, and Margret Sauter. 2011. "Phytosulfokine-α Controls Hypocotyl Length and Cell Expansion in Arabidopsis Thaliana through Phytosulfokine Receptor 1." *PLoS ONE* 6(6):e21054.

Sun, Jia-Qiang, Hong-Ling Jiang, and Chuan-You Li. 2011. "Systemin/Jasmonate-Mediated Systemic Defense Signaling in Tomato." *Molecular Plant* 4(4):607–15.

Tabata, Ryo and Shinichiro Sawa. 2014. "Maturation Processes and Structures of Small Secreted Peptides in Plants." *Frontiers in Plant Science* 5(July):311.

Taiz, Lincoln and Eduardo Zeiger. 2010. *Plant Physiology.*

Takahashi, Fuminori and Kazuo Shinozaki. 2019. "Long-Distance Signaling in Plant Stress Response." *Current Opinion in Plant Biology* 47:106–11.

Thain, J. F., I. R. Gubb, and D. C. Wildon. 1995. "Depolarization of Tomato Leaf Cells by Oligogalacturonide Elicitors." *Plant, Cell & Environment* 18(2):211–14.

Theodoulou, Frederica L., Michael Holdsworth, and Alison Baker. 2006. "Peroxisomal ABC Transporters." *FEBS Letters* 580(4):1139–55.

Truitt, C. L., Han-Xun Wei, and Paul Pare. 2004. "A Plasma Membrane Protein from Zea Mays Binds with the Herbivore Elicitor Volicitin." *THE PLANT CELL* 16(2):523–32.

Unsicker, Sybille B., Grit Kunert, and Jonathan Gershenzon. 2009. "Protective Perfumes: The Role of Vegetative Volatiles in Plant Defense against Herbivores." *Current Opinion in Plant Biology* 12(4):479–85.

Vidhyasekaran, P. 2015. "Salicylic Acid Signaling in Plant Innate Immunity." Pp. 27–122 in *Plant Hormone Signaling Systems in Plant Innate Immunity, Signaling and Communication in Plants*. Dordrecht: Springer Netherlands.

Wan, J., K. Tanaka, X. C. Zhang, G. H. Son, L. Brechenmacher, T. H. N. Nguyen, and G. Stacey. 2012. "LYK4, a Lysin Motif Receptor-Like Kinase, Is Important for Chitin Signaling and Plant Innate Immunity in Arabidopsis." *PLANT PHYSIOLOGY* 160(1):396–406.

Wang, Jiehua and C. Peter Constabel. 2004. "Polyphenol Oxidase Overexpression in Transgenic Populus Enhances Resistance to Herbivory by Forest Tent Caterpillar (Malacosoma Disstria)." *Planta* 220(1):87–96.

Wang, Kevin L. C., Hai Li, and Joseph R. Ecker. 2002. "Ethylene Biosynthesis and Signaling Networks." *The Plant Cell* 14(suppl 1):S131–51.

Wang, Lei, Elias Einig, Marilia Almeida-Trapp, Markus Albert, Judith Fliegmann, Axel Mithöfer, Hubert Kalbacher, and Georg Felix. 2018. "The Systemin Receptor SYR1 Enhances Resistance of Tomato against Herbivorous Insects." *Nature Plants* 4(3):152–56.

Wang, Xiaofeng, Michael B. Goshe, Erik J. Soderblom, Brett S. Phinney, Jason A. Kuchar, Jia Li, Tadao Asami, Shigeo Yoshida, Steven C. Huber, and Steven D. Clouse. 2005. "Identification and Functional Analysis of in Vivo Phosphorylation Sites of the Arabidopsis BRASSINOSTEROID-INSENSITIVE1 Receptor Kinase." *The Plant Cell* 17(June):1685–1703.

Wang, Xiaofeng, Uma Kota, Kai He, Kevin Blackburn, Jia Li, Michael B. Goshe, Steven C. Huber, and Steven D. Clouse. 2008. "Sequential Transphosphorylation of the BRI1/BAK1 Receptor Kinase Complex Impacts Early Events in Brassinosteroid Signaling." *Developmental Cell* 15(2):220–35.

Wang, Yin, Maxime Chantreau, Richard Sibout, and Simon Hawkins. 2013. "Plant Cell Wall Lignification and Monolignol Metabolism." *Frontiers in Plant Science* 4(July):Article 220.

War, Abdul Rasheed, Gaurav Kumar Taggar, Barkat Hussain, Monica Sachdeva Taggar, Ramakrishnan M. Nair, and Hari C. Sharma. 2018. "Plant Defense Against Herbivory and Insect Adaptations." *AoB PLANTS* 10(4):1–19.

Wasternack, Claus, Irene Stenzel, Bettina Hause, Gerd Hause, Claudia Kutter, Helmut Maucher, Jana Neumerkel, Ivo Feussner, and Otto Miersch. 2006. "The Wound Response in Tomato – Role of Jasmonic Acid." *Journal of Plant Physiology* 163(3):297–306.

Watanabe, T. and S. Sakai. 1998. "Effects of Active Oxygen Species and Methyl Jasmonate on Expression of the Gene for a Wound-Inducible 1-Aminocyclopropane-1-Carboxylate Synthase in Winter Squash (Cucurbita Maxima)." *Planta* 206(4):570–76.

Wera, Stefaan and B. A. Hemmings. 1995. "Serine/Threonine Protein Phosphatases." *Biochemical Journal* 311(1):17–29.

Whitfield, W. G., M. A. Chaplin, K. Oegema, H. Parry, and D. M. Glover. 1995. "The 190 KDa Centrosome-Associated Protein of Drosophila Melanogaster Contains Four Zinc Finger Motifs and Binds to Specific Sites on Polytene Chromosomes." *Journal of Cell Science* 108:3377–87.

van Wijk, Klaas J., Giulia Friso, Dirk Walther, and Waltraud X. Schulze. 2014. "Meta-Analysis of Arabidopsis Thaliana Phospho-Proteomics Data Reveals Compartmentalization of Phosphorylation Motifs." *The Plant Cell* 26(June):2367–89.

Wu, Xu Na, Clara Sanchez Rodriguez, Heidi Pertl-Obermeyer, Gerhard Obermeyer, and Waltraud X. Schulze. 2013. "Sucrose-Induced Receptor Kinase SIRK1 Regulates a Plasma Membrane Aquaporin in Arabidopsis." *Molecular & Cellular Proteomics* 12(10):2856–73.

Wu, Xu Na, Lin Xi, Heidi Pertl-Obermeyer, Zhi Li, Liang-Cui Chu, and Waltraud X. Schulze. 2017. "Highly Efficient Single-Step Enrichment of Low Abundance Phosphopeptides from Plant Membrane Preparations." *Frontiers in Plant Science* 8(September):1–14.

Xing, Hui-Li, Li Dong, Zhi-Ping Wang, Hai-Yan Zhang, Chun-Yan Han, Bing Liu, Xue-Chen Wang, and Qi-Jun Chen. 2014. "A CRISPR/Cas9 Toolkit for Multiplex Genome Editing in Plants." *BMC Plant Biology* 14(1):327.

Xing, Tim and André Laroche. 2011. "Revealing Plant Defense Signaling: Getting More Sophisticated with Phosphoproteomics." *Plant Signaling & Behavior* 6(10):1469–74.

Xu, Siming, Chao-Jan Liao, Namrata Jaiswal, Sanghun Lee, Dae-Jin Yun, Sang Yeol Lee, Michael Garvey, Ian Kaplan, and Tesfaye Mengiste. 2018. "Tomato PEPR1 ORTHOLOG RECEPTOR-LIKE KINASE1 Regulates Responses to Systemin, Necrotrophic Fungi, and Insect Herbivory." *The Plant Cell* 30(9):2214–29.

Xu, Ting Ting, Shi Chao Ren, Xiu Fen Song, and Chun Ming Liu. 2015. "CLE19 Expressed in the Embryo Regulates Both Cotyledon Establishment and Endosperm Development in Arabidopsis." *Journal of Experimental Botany* 66(17):5217–27.

Yamaguchi, Yube and Alisa Huffaker. 2011. "Endogenous Peptide Elicitors in Higher Plants." *Current Opinion in Plant Biology* 14(4):351–57.

Yan, Liuhua, Qingzhe Zhai, Jianing Wei, Shuyu Li, Bao Wang, Tingting Huang, Minmin Du, Jiaqiang Sun, Le Kang, Chang-Bao Li, and Chuanyou Li. 2013. "Role of Tomato Lipoxygenase D in Wound-Induced Jasmonate Biosynthesis and Plant Immunity to Insect Herbivores" edited by H. Yu. *PLoS Genetics* 9(12):e1003964.

Yazaki, Kazufumi, Gen Ichiro Arimura, and Toshiyuki Ohnishi. 2017. "'Hidden' Terpenoids in Plants: Their Biosynthesis, Localization and Ecological Roles." *Plant and Cell Physiology* 58(10):1615–21.

Yu, Lita P., Andrea K. Miller, and Steven E. Clark. 2003. "POLTERGEIST Encodes a Protein Phosphatase 2C That Regulates CLAVATA Pathways Controlling Stem Cell Identity at Arabidopsis Shoot and Flower Meristems." *Current Biology* 13(3):179–88.

Zauber, Henrik and Waltraud X. Schulze. 2012. "Proteomics Wants CRacker: Automated Standardized Data Analysis of LC-MS Derived Proteomic Data." *Journal of Proteome Research* 11(11):5548–55.

Zhai, Qingzhe, Chun Yan, Lin Li, Daoxin Xie, and Chuanyou Li. 2017. "Jasmonates." Pp. 243–72 in *Hormone Metabolism and Signaling in Plants*, edited by J. Li, C. Li, and S. Smith. Elsevier.

Zhang, Dingbo, Huawei Zhang, Tingdong Li, Kunling Chen, Jin-Long Qiu, and Caixia Gao. 2017. "Perfectly Matched 20-Nucleotide Guide RNA Sequences Enable Robust Genome Editing Using High-Fidelity SpCas9 Nucleases." *Genome Biology* 18(1):191.

Zhang, Haiyan, Pengli Yu, Jiuhai Zhao, Hongling Jiang, Haiyang Wang, Yingfang Zhu, Miguel A. Botella, Jozef Šamaj, Chuanyou Li, and Jinxing Lin. 2018a. "Expression of Tomato Prosystemin Gene in Arabidopsis Reveals Systemic Translocation of Its MRNA and Confers Necrotrophic Fungal Resistance." *New Phytologist* 217(2):799–812.

Zhang, Huan, Zhangjian Hu, Cui Lei, Chenfei Zheng, Jiao Wang, Shujun Shao, Xin Li, Xiaojian Xia, Xinzhong Cai, Jie Zhou, Yanhong Zhou, Jingquan Yu, Christine H. Foyer, and Kai Shi. 2018b. "A Plant Phytosulfokine Peptide Initiates Auxin-Dependent Immunity through Cytosolic Ca 2+ Signaling in Tomato." *The Plant Cell* 30(3):652–67.

Zhang, Jie, Wei Li, Tingting Xiang, Zixu Liu, Kristin Laluk, Xiaojun Ding, Yan Zou, Minghui Gao, Xiaojuan Zhang, She Chen, Tesfaye Mengiste, Yuelin Zhang, and Jian Min Zhou. 2010. "Receptor-like Cytoplasmic Kinases Integrate Signaling from Multiple Plant Immune Receptors and Are Targeted by a Pseudomonas Syringae Effector." *Cell Host and Microbe* 7(4):290–301.

Zhang, Tong, Sixue Chen, Haiying Li, Cecilia Silva-Sanchez, and Jinna Li. 2015. "Phosphoproteomics Technologies and Applications in Plant Biology Research." *Frontiers in Plant Science* 6(June):1–9.

Zhang, Yan, Hye Kyong Kweon, Christian Shively, Anuj Kumar, and Philip C. Andrews. 2013. "Towards Systematic Discovery of Signaling Networks in Budding Yeast Filamentous Growth Stress Response Using Interventional Phosphorylation Data" edited by C. A. Orengo. *PLoS Computational Biology* 9(6):e1003077.

Zheng, Ting, Yingzi Hou, Pingjing Zhang, Zhenxi Zhang, Ying Xu, Letian Zhang, Leilei Niu, Ying Yang, Da Liang, Fan Yi, Wei Peng, Wenjian Feng, Ying Yang, Jianxin Chen, York Yuanyuan Zhu, Li-He Zhang, and Quan Du. 2017. "Profiling Single-Guide RNA Specificity Reveals a Mismatch Sensitive Core Sequence." *Scientific Reports* 7(1):40638.

Zhu-Salzman, Keyan, Dawn S. Luthe, and Gary W. Felton. 2008. "Arthropod-Inducible Proteins: Broad Spectrum Defenses against Multiple Herbivores." *PLANT PHYSIOLOGY* 146(3):852–58.

Zhu, Xiaoxiao, Yajie Xu, Shanshan Yu, Lu Lu, Mingqin Ding, Jing Cheng, Guoxu Song, Xing Gao, Liangming Yao, Dongdong Fan, Shu Meng, Xuewen Zhang, Shengdi Hu, and Yong Tian. 2014. "An Efficient Genotyping Method for Genome-Modified Animals and Human Cells Generated with CRISPR/Cas9 System." *Scientific Reports* 4(1):6420.

Ziegler, Jörg, Irene Stenzel, Bettina Hause, Helmut Maucher, Mats Hamberg, Rudi Grimm, Martin Ganal, and Claus Wasternack. 2000. "Molecular Cloning of Allene Oxide Cyclase: The Enzyme Establishing the Stereochemistry of Octadecanoids and Jasmonates." *Journal of Biological Chemistry* 275(25):19132–38.

6. Appendix

6.1. Appendix A.1 Phosphopeptides List

The list of the quantified phosphopeptides including normalized ion intensities, standard deviation and number of spectra is available in the enclosed CD.

6.2. Appendix A.2 *k*-means Clusters

The list for *k*-means clusters for all identified phosphopeptides including A17 response shift is available in the enclosed CD.

6.3. Appendix A.3 Overrepresented Systemin-specific Functional Categories

Functional categories (MAPMAN bins) significantly overrepresented in k-means clusters of Systemin-specific proteins. The p-values of Fisher's Exact test for each category is shown. Numbers represent the percentage of the phosphopeptides in each functional category per cluster. Color range from green to red represents percentage range from high to low values.

p-value	BIN code	MAPMAN	Sys cluster A (0min)	Sys cluster B (2min)	Sys cluster C (5min)	Sys cluster D (15min)	Sys cluster F
3,74E-19	30.1	signalling.in sugar and nutrient physiology	0,9				0,3
	30.11	signalling.light	0,9		0,7	1,4	0,8
	30.2	signalling.receptor kinases	3,9	2,8	0,7	3,1	4,4
	30.3	signalling.calcium	3,0	1,7	4,7	3,1	3,9
	30.4	signalling.phosphinositides	0,3	0,6	0,7	1,1	0,5
	30.5	signalling.G-proteins	3,9	2,8	3,3	5,6	3,1
	30.6	signalling.MAP kinases	0,3			0,3	
	30.99	signalling.unspecified	0,3	1,1	0,7	0,3	0,3
2,323E-36	34	transport		0,6		0,3	
	34.1	transport.p- and v-ATPases	0,6	1,1	1,3	1,7	1,3
	34.10	transport.nucleotides					0,3
	34.12	transport.metal	0,6	1,7		0,8	1,0
	34.13	transport.peptides and oligopeptides	0,6		0,7	0,6	0,3
	34.14	transport.unspecified cations	0,9	2,3	2,7	0,6	1,3
	34.15	transport.potassium	1,5	1,1	2,7	1,1	1,5
	34.16	transport.ABC transporters and multidrug resistance systems	2,4	2,3	4,0	1,4	1,8
	34.18	transport.unspecified anions	1,2	1,7	3,3	0,3	1,3
	34.19	transport.Major Intrinsic Proteins	1,2	0,6	1,3	0,6	1,5
	34.2	transport.sugars	1,5	2,8	0,7	1,4	1,3
	34.21	transport.calcium		1,1	1,3	0,6	0,5
	34.3	transport.amino acids	0,9	1,1	2,0	0,6	1,0
	34.5	transport.ammonium					0,3
	34.7	transport.phosphate		0,6			

Appendix

p-value	bin	description					
	34.8	transport.metabolite transporters at the envelope membrane	0,3	0,6	0,7	0,6	0,8
	34.9	transport.metabolite transporters at the mitochondrial membrane	0,3				0,3
	34.98	transport.membrane system unknown	0,3				0,3
	34.99	transport.misc	2,4		2,0	1,7	2,1
0,017	20	stress		0,6	0,7		
	20.1	stress.biotic	1,2	2,8	3,3	1,7	2,6
	20.2	stress.abiotic	3,0	1,1	0,7	2,5	2,6
0,028	29.1	protein.aa activation		0,6	0,7	0,3	0,3
	29.2	protein.synthesis	3,6		2,0	4,2	2,3
	29.3	protein.targeting	2,4	4,5	4,0	1,7	3,4
	29.4	protein.postranslational modification	13,8	14,2	12,7	16,4	15,2
	29.5	protein.degradation	8,1	8,5	13,3	6,9	8,5
	29.6	protein.folding	0,3	0,6	0,7	0,3	0,5
	29.7	protein.glycosylation		0,6	0,7		0,5
3,095E-16	31.1	cell.organisation	2,4	3,4	2,0	3,1	4,1
	31.2	cell.division	0,9	0,6		1,1	0,5
	31.3	cell.cycle	1,8	1,7	0,7	1,1	1,3
	31.4	cell.vesicle transport	5,7	7,4	8,0	4,7	5,7
	31.5	cell.cell death	0,3	0,6			
8,456E-06	27.1	RNA.processing	3,9	2,8	2,7	4,7	2,3
	27.2	RNA.transcription				0,6	0,3
	27.3	RNA.regulation of transcriptionfamily	12,6	9,7	5,3	13,1	11,1
	27.4	RNA.RNA binding	6,6	5,1	2,0	5,3	2,6
0,00003938	10.1	cell wall.precursor synthesis	0,9	0,6	1,3	1,1	0,5
	10.2	cell wall.cellulose synthesis	1,2	4,0	2,0	1,1	1,0
	10.3	cell wall.hemicellulose synthesis					0,3
	10.4	cell wall.pectin synthesis				0,3	
	10.5	cell wall.cell wall proteins		0,6		0,3	0,3
	10.6	cell wall.degradation	0,3		1,3	0,3	0,3
	10.8	cell wall.pectin*esterases		0,6			0,5
	17.1.1	abscisic acid.synthesis-degradation	0,6	0,6	0,7	0,6	0,5
	17.1.2	abscisic acid.signal transduction	0,3				

17.1.3	abscisic acid.induced-regulated-responsive-activated				0,3	
17.2.2	auxin.signal transduction	0,3		0,7	0,3	1,0
17.2.3	auxin.induced-regulated-responsive-activated	0,6	1,7	0,7	1,1	1,3
17.4.2	hormone metabolism.cytokinin. signal transduction					0,3
17.5.1.1	ethylene.synthesis-degradation.1-aminocyclopropane-1-carboxylate synthase		0,6		0,6	
17.5.2	ethylene.signal transduction	0,6		0,7		0,3
17.8.1	hormone metabolism.salicylic acid.synthesis-degradation					0,3

6.4. Appendix A.4 Distribution of Systemin-Specific Overrepresented Functional Categories Phosphopeptides in the A17 Response Shift Groups

A list showing the percentage distribution of Systemin-specific overrepresented functional categories phosphopeptides in the A17 response shift groups. The p-values of Fisher's Exact test for each category are shown. The numbers represent the percentage of the phosphopeptides in each of A17 response shift group. Color range from green to red represents percentage range from high to low values.

p-value	BIN code	MAPMAN bin	NOT in A17	A17 late	A17 early	A17 equal
3,57E-05	30.2	signalling.receptor kinases	3,1	4,4	4,2	1,4
	30.3	signalling.calcium	3,6	3,3	3,2	3,8
	30.4	signalling.phosphinositides	1,0	0,4	1,1	0,5
	30.5	signalling.G-proteins	4,1	3,9	4,2	2,8
	30.6	signalling.MAP kinases	0,5	0,2		0,5
	30.99	signalling.unspecified		0,4	0,4	0,5
0,0025	34	transport		0,2	0,2	
	34.1	transport.p- and v-ATPases	2,6	0,7	1,5	
	34.10	transport.nucleotides			0,2	
	34.12	transport.metal	1,5	0,2	0,6	1,4
	34.13	transport.peptides and oligopeptides		1,1	0,2	0,5
	34.14	transport.unspecified cations	0,5	1,1	1,1	1,4
	34.15	transport.potassium	1,5	1,1	1,3	0,9
	34.16	transport.ABC transporters and multidrug resistance systems	2,1	2,4	2,1	0,9
	34.18	transport.unspecified anions	1,0	1,3	1,1	0,9
	34.19	transport.Major Intrinsic Proteins	0,5	1,1	1,7	
	34.2	transport.sugars	1,0	1,3	1,7	0,9
	34.21	transport.calcium	0,5	0,4	0,4	0,9
	34.3	transport.amino acids	1,0	1,3	0,8	0,5
	34.7	transport.phosphate	0,5			
	34.8	transport.metabolite transporters at the envelope membrane	0,5	0,2	0,6	1,4
	34.9	transport.metabolite transporters at the mitochondrial membrane		0,2	0,2	
	34.98	transport.membrane system		0,2	0,2	0,5

Appendix

p-value	code		V1	V2	V3	V4
		unknown				
	34.99	transport.misc	1,0	2,0	2,1	1,9
	20	stress	0,5		0,2	
	20.1	stress.biotic	3,1	2,0	2,1	3,8
	20.2	stress.abiotic	3,1	2,2	2,7	3,8
0,0438	29.1	protein.aa activation	0,5	0,2	0,2	0,5
	29.2	protein.synthesis	2,6	3,5	2,1	3,8
	29.3	protein.targeting	3,1	2,6	2,7	2,4
	29.4	protein.postranslational modification	10,8	15,7	16,2	15,1
	29.5	protein.degradation	9,3	8,3	8,0	7,1
	29.6	protein.folding	0,5	0,2	0,2	0,5
	29.7	protein.glycosylation		0,2	0,4	0,5
1,43E-06	31.1	cell.organisation	3,6	2,6	4,0	2,8
	31.2	cell.division	1,0	0,9	0,8	0,5
	31.3	cell.cycle	1,5	1,3	1,3	1,4
	31.4	cell.vesicle transport	7,7	4,8	5,9	6,6
	31.5	cell.cell death	0,5	0,2		
0,0463	27.1	RNA.processing	3,6	3,3	3,8	5,2
	27.2	RNA.transcription	0,5	0,4	0,2	0,5
	27.3	RNA.regulation of transcription	10,8	12,2	9,9	11,8
	27.4	RNA.RNA binding	2,1	4,8	3,6	5,2
	10.1	cell wall.precursor synthesis	0,5	0,9	0,8	0,5
	10.2	cell wall.cellulose synthesis	2,1	1,5	0,8	2,4
	10.3	cell wall.hemicellulose synthesis			0,2	
0,00702	10.4	cell wall.pectin synthesis		0,2		
	10.5	cell wall.cell wall proteins	0,5	0,2	0,2	0,5
	10.6	cell wall.degradation	0,5	0,4	0,2	
	10.8	cell wall.pectin*esterases			0,6	
	17.1.1	abscisic acid.synthesis-degradation	1,0	0,4		0,5
	17.1.2	abscisic acid.signal transduction		0,2		
	17.1.3	abscisic acid.induced-regulated-responsive-activated			0,2	0,5
	17.2.2	auxin.signal transduction	0,5	0,2	0,8	
	17.2.3	auxin.induced-regulated-responsive-activated	2,1	0,9	1,1	0,5
	17.4.2	cytokinin.signal transduction	0,5			
	17.5.1.1	ethylene.synthesis-degradation.1-aminocyclopropane-1-carboxylate synthase		0,2		
	17.5.2	ethylene.signal transduction		0,7	0,4	
	17.8.1	hormone metabolism.salicylic acid.synthesis-degradation			0,2	

6.5. Appendix A.5 List of Systemin-specifically Induced Kinases and Phosphatases

The list of the Systemin-specifically induced Kinases and Phosphatases is available in the enclosed CD. Included are the accession numbers for these proteins, the cluster they were attributed to after Systemin, A17 and water treatment, the shift in cluster after A17 and water as compared to Systemin treatment, their annotation, and the closest *Arabidopsis* homologs.

6.6. Appendix A.6 Pearson Correlation Analysis for Systemin-Specific Kinases and Phosphatases

Pearson correlation analysis between Systemin-specific kinases and phosphatases and the proteins from Systemin-induced functional categories. The file on the enclosed CD contains two data sheets. One is listing the proteins and their phosphopeptides used in the Pearson correlation analysis. The other data sheet lists the results of the Pearson correlation analysis network.

6.7. Appendix A.7 Mass Spectrometry Analysis of *in vitro* Kinase and Phosphatase Assays

(**A**) Representative spectrum of kinase reaction with recombinant MPK2 resulting in phosphopeptide GLDIETIQQSY(pT)V. (**B**) Representative spectrum of phosphatase reaction with recombinant PLL5 phosphatase resulting in dephosphorylated peptide GLDIETIQQSYTV. (**C**) Summary of ion intensities of phosphorylated and dephosphorylated C-terminal peptide of H^+-ATPase LHA1. Input peptides are shown as bars without patterns, reaction results are indicated as hatched bars. Averages with standard deviations of two biological replicates from two independent recombinant protein preparations are shown.

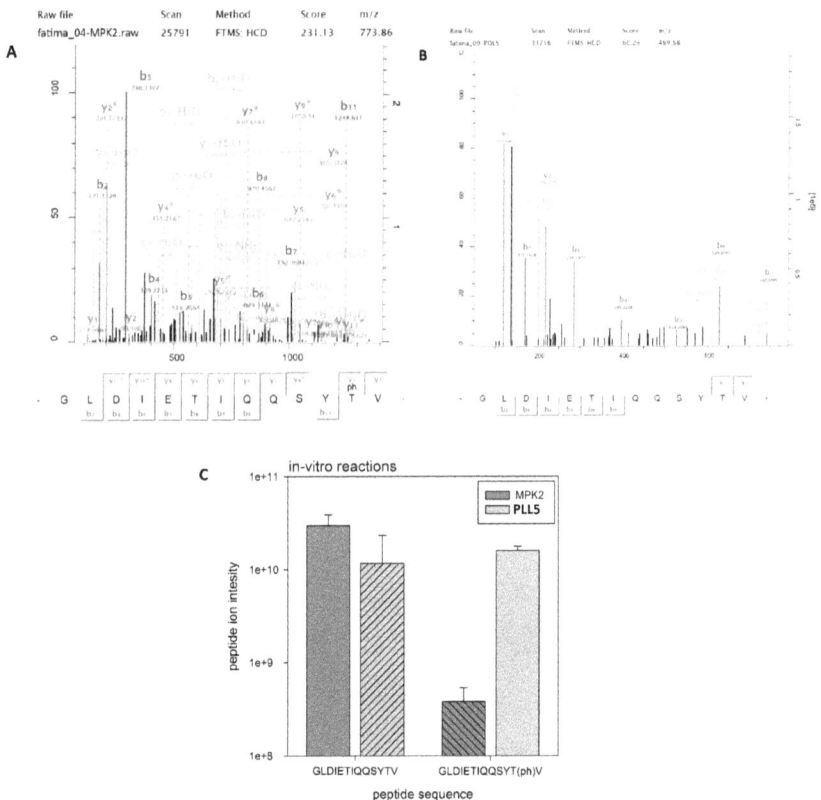

6.8. Appendix A.8 List of all RLKs Identified in the Data Set

List of all RLKs identified in the data set. The file on the enclosed CD lists the RLK accessions, their identified phosphopeptides, and the cluster they were attributed to. Also indicated for each RLK is whether or not they were identified in response to Flg22 and xylanase treatments according to Benschop et al. (2007); Mattei et al. (2016); Nühse et al. (2007); Rayapuram et al. (2014), and whether or not their expression is upregulated after wounding.

6.9. Appendix A.9 List of Mutations Identified in the RLK KO *S. peruvianum* Cell Culture

A list of mutations identified in the cell suspension cultures of each RLK CRISPR/Cas9 construct. The mutations identified at the 1st and 2nd sgRNA positions are listed as well as the effect of these mutations on the protein sequence of the RLK. (+) indicates an insertion mutation. (–) indicates a deletion. For big deletions, the position of the deletion relative to the start codon is indicated in brackets. The deleted residues in the protein sequence also are indicated in brackets. a.as: amino acids. WT: wild-type sequence, meaning that no mutations were detected.

CRISPR/Cas9 construct	Mutations at 1st sgRNA position	Mutations at 2nd sgRNA position	Effect of mutation on protein sequence
Solyc01g109650 (LRRXIV)	+1nt	WT	A truncated protein of 71a.as
	-1nt	WT	A truncated protein of 97a.as
	-6nt	-16nt (236-251nt)	A truncated protein of 90a.as
	-4nt	WT	A truncated protein of 64a.as
		-52nt (199-250nt)	A truncated protein of 74a.as
	-4nt	+1nt	A truncated protein of 64a.as
	-1nt	+1nt	A truncated protein of 97a.as
		-93nt (159-251nt)	A protein shorter with 31a.as (55-85a.a)
	-3nt	-1nt	A truncated protein of 96a.as
Solyc02g070000 (GHR1)	+1nt	+1nt	A truncated protein of 151a.as
	+1nt	-6nt	A truncated protein of 151a.as
		-115nt (452-566nt)	A truncated protein of 134a.as
Solyc07g063000 (PSKR2)		-104nt (472-575nt)	A truncated protein of 97a.as

	WT	+1nt	A truncated protein of 132a.as
	-3nt	WT	A truncated protein of 64a.as
	-9nt	-5nt	A truncated protein of 127a.as
	+1nt	-9nt	A truncated protein of 73a.as
	+1nt	-2nt	A truncated protein of 73a.as
	+1nt	+1nt	A truncated protein of 73a.as
	-4nt	-7nt	A truncated protein of 79a.as
	-4nt	-75nt (256-330nt)	A truncated protein of 79a.as
	-2nt	WT	A truncated protein of 72a.as
Solyc09g083210 (LYK4)	-1nt	+2nt	A truncated protein of 158a.as
	+1nt	-97nt (530-626nt)	A truncated protein of 158a.as
	-1nt	WT	A truncated protein of 158a.as
	+1nt	-2nt	A truncated protein of 103a.as
	-164nt (235-398nt)	-3nt	A truncated protein of 79a.as
	-164nt (235-398nt)	WT	A truncated protein of 79a.as
	-39nt (262-300nt)	+1nt	A truncated protein of 199a.as
	WT	-15nt (601-615nt)	A protein shorter with -5aa (201-205a.a)
Solyc12g036330 (LRK10L1.2)	-5nt	-224nt (318-541nt)	A truncated protein of 60a.as
	-2nt	-1nt	A truncated protein of 61a.as

	-2nt	+1nt	A truncated protein of 61a.as
		-303nt (166-468nt)	A protein shorter with 101a.as (56-156a.a)
	-3nt	-16nt (452-467nt)	A truncated protein of 181a.as
	-3nt	-1nt	A truncated protein of 186a.as
	-9nt	-1nt	A truncated protein of 184a.as
	+1nt	-1nt	A truncated protein of 62a.as
	-1nt	-1nt	A truncated protein of 173a.as
	-23nt (131-153nt)		
	-18nt (159-176nt)		
	-35nt (163-197nt)	-1nt	A truncated protein of 57a.as
Solyc03g082470 (SYR1)	+1nt	-8nt	A truncated protein of 100a.as
	-1nt	-8nt	A truncated protein of 106a.as
	-3nt	-1nt	A truncated protein of 427a.as
	-7nt	-19nt	A truncated protein of 104a.as
	-7nt	-54nt (1230-1283nt)	A truncated protein of 104a.as
	-6nt	-2nt	A truncated protein of 426a.as
		-964nt (290-1253nt)	A truncated protein of 96a.as
Solyc03g082450 (SYR2)	-1nt	+1nt	A truncated protein of 140a.as
	+1nt	-1nt	A truncated protein of 107a.as
	+2nt	-2nt	A truncated protein of 141a.as

	+1nt	-5nt	A truncated protein of 107a.as
	+1nt	+1nt	A truncated protein of 107a.as
	-4nt	+1nt	A truncated protein of 139a.as
	+2nt	-5nt	A truncated protein of 141a.as
	-1nt	5 Substitution	A truncated protein of 196a.as
	-1nt	-1nt	A truncated protein of 140a.as
	-1nt	-8nt	A truncated protein of 140a.as
	-6nt	-1nt	A truncated protein of 195a.as
	-6nt	-196nt (463-658nt)	A truncated protein of 161a.as
Solyc03g123860 (PORK1)	1st and 2nd sgRNAs were WT	3rd sgRNA: -4nt	A truncated protein of 596a.as
	1st and 2nd sgRNAs were WT	3rd sgRNA: +1nt	A truncated protein of 601a.as
	1st and 2nd sgRNAs were WT	3rd sgRNA: -1nt	A truncated protein of 597a.as
	1st and 2nd sgRNAs were WT	3rd sgRNA: -35nt	A truncated protein of 589a.as

6.10. Appendix A.10 List of Mutations Identified in T_1 RLK Mutated Tomato Lines

A list of mutations identified in T_1 RLK mutated lines, which their second generation plants were used for the wound experiment. The mutations identified at the 1st and 2nd sgRNA positions are listed as well as the effect of these mutations on the protein sequence of the RLK. + indicates an insertion mutation. – indicates a deletion mutation. For big deletions, the position of the deletion relative to the start codon is indicated in brackets. The deleted residues in the protein sequence also are indicated in brackets. a.as: amino acids. WT: wild-type sequence, meaning that no mutations were detected.

RLK CRISPR/Cas9 construct	T_1 line	Mutations at 1st sgRNA position	Mutations at 2nd sgRNA position	Effect of mutation on protein sequence
Solyc01g109650 (LRRXIV)	lrrxiv.2	-1nt	-16nt (235-250nt)	A truncated protein of 86a.as
	lrrxiv.2	-6nt	-8nt	A truncated protein of 87a.as
Solyc02g070000 (GHR1)	ghr1.2	-2nt	-2nt	A truncated protein of 150a.as
	ghr1.6	-2nt	-4nt	A truncated protein of 150a.as
Solyc07g063000 (PSKR2)	pskr2.2	+1nt	WT	A truncated protein of 73a.as
Solyc09g083210 (LYK4)	lyk4.1	-6nt	+1nt	A truncated protein of 210a.as
	lyk4.2	-5nt	-15nt (602-616nt)	A truncated protein of 101a.as
Solyc12g036330 (LRK10L1.2)	lrk10l1.2.1	-2nt	+1nt	A truncated protein of 61a.as
Solyc03g082470 (SYR1)	syr.1	-1nt	+1nt	A truncated protein of 106a.as
Solyc03g082450 (SYR2)	syr2.3	-1nt	WT	A truncated protein of 140a.as
	syr2.6	-1nt	WT	A truncated protein of 140a.as

7. Declaration

Affidavit according to Sec. 7(7) of the University of Hohenheim doctoral degree regulations for Dr. rer. nat.

1. For the dissertation submitted on the topic

"Phosphoproteomics Analysis of the Systemin Signaling Pathway in Tomato" I hereby declare that I independently completed the work.

2. I only used the sources and aids documented and only made use of permissible assistance by third parties. In particular, I properly documented any contents which I used - either by directly quoting or paraphrasing - from other works.

3. I did not accept any assistance from a commercial doctoral agency or consulting firm.

4. I am aware of the meaning of this affidavit and the criminal penalties of an incorrect or incomplete affidavit.

I hereby confirm the correctness of the above declaration: I hereby affirm in lieu of oath that I have, to the best of my knowledge, declared nothing but the truth and have not omitted any information.

_____   _____

Stuttgart, 08.05.2019   Fatima Haj Ahmad

8. Acknowledgement

I would like to express my deep gratitude to my supervisor Prof. Dr. Andreas Schaller for giving me the chance of carrying my PhD research in his institute. His guidance, patience and trust in all time of this project helped me to acquire the expertise that will enlighten the path of my future career.

Besides my supervisor, I would like to thank sincerely Prof. Dr. Waltraud Schulze for her friendly willingness to finish reviewing and evaluating this thesis in a short time. And I wish to deeply thank her for her support and help during the phosphoproteomics data analysis and writing the paper that emerged from that. My sincere thanks go also to Prof. Dr. Artur Pfitzner as a member of the examination committee.

Special thanks go to Dr. Annick Stintzi for her support, motivation and guidance at the professional and personal levels. I totally appreciate her fruitful discussions and advices that allowed me to develop my skills, which will help me to face my future challenges.

I would like to thank the German Academic Exchange Service for supporting me financially during my research at the University of Hohenheim.

The assistance and worthful discussions given by the members of Plant Systems Biology Institute during membrane fraction proteins extraction, chimera receptors expression and *in vitro* assays are greatly appreciated. I would like to thank also all my colleagues in our institute for their friendship, great work environment and useful ideas and suggestions.

I wish to acknowledge the help provided by all members of TA team in the institute as well as the green house team. Thank you for your assistance in all the enormous work we had. Without your professional support most of this work could not be done.

My deep thanks go to Prof. Dr. Ghandi Anfoka in Jordan. The experience I gained in his laboratory gave me more confidence to face the first challenges I had in this project. My special thanks are extended also to Plamper family for their generosity and kindness, who opened their house for us during our stay in Köln and gave us a chance to experience the German culture and language from the very beginning of our time in Germany.

From the bottom of my heart I would like to thank my beloved family especially my husband Hamzeh Al Khalaileh, who always believed in me, for his endless love, support and patience during all the hard times we faced in this journey. Without your help and encouragement, I would not be able to accomplish all of this. To my parents in Jordan thank you that you are always there for me. To both of you this work is dedicated…

9. Curriculum Vitae

Personal Information

- Date and place of Birth: 17th of Oct. 1982, Amman
- Marital Status: Married, 2 Children
- Nationality: Jordanian
- E-Mail: fatimahajahmad@yahoo.com

Education

06/2012 – 5/2019	• **University of Hohenheim, Institute of Plant Physiology and Biotechnology (Stuttgart/Germany)** *PhD Student* - Thesis topic: Phosphoproteomics analysis of the Systemin Signaling Pathway in Tomato. Thesis Defense in Sommer 2019
03/2004 – 02/2008	• **Al-Balqa' Applied University (Al-Salt/Jordan) M. Sc. in Biotechnology** - Thesis Topic: The Effect of *Cucumber mosaic virus*, *Potato virus Y* and *Tobacco mosaic virus* Infection on the Gene Silencing Triggered against *Tomato yellow leaf curl virus*
10/2000 – 02/2004	• **Jordan University of Science and Technology (Irbid/Jordan) B. Sc. in Biotechnology and Genetic Engineering**
06/2000	• **Wadi Al-Seer Secondary School (Amman/Jordan) High School certificate**

Professional Experience

06/2015 – 06/2019	• **University of Hohenheim, Institute of Plant Physiology and Biotechnology (Stuttgart/Germany)** *DAAD (German Academic Exchange Service) scholarship holder*

06/2009 – 05/2015

- **Al-Balqa' Applied University, Faculty of Agricultural Technology, Department of Biotechnology (Al-Salt/Jordan)**
  Laboratory supervisor

06/2006 – 05/2015

- **Al-Balqa' Applied University, Faculty of Agricultural Technology, Department of Biotechnology (Al-Salt/Jordan)**
  Research Assistant

Publications

2016

- Anfoka, G., Al-Talb, M. and **Haj-Ahmad, F.** 2016. A new isolate of Tomato Yellow Leaf Curl Axarquia Virus associated with tomato yellow leaf curl disease in Jordan. Journal of Plant Pathology 98: 145-149.

2014

- Lapidot, M., Gelbart, D., Gal-On, A., Anfoka, G., **Haj Ahmed, F.**, Abou-Jawada, Y., Sobh, H., Mazyad, H., Aboul-Ata, A., El-Attar, A., Ali-Shtayeh, M., Jamous, R., Polston, J. and Duffy, S. 2014. Frequent migration of introduced cucurbit-infecting begomoviruses among Middle Eastern countries. Virology Journal 11: 181

2014

- Anfoka, G., **Haj Ahmad, F.**, Altaleb, M., Abadi, M., Abubaker, S., Levy D., Rosner, A., Czosnek, H. 2014. First report of recombinant potato virus Y strain infecting potato in Jordan. Plant Disease 98 (7): 1017.3.

2014

- Anfoka, G., **Haj Ahmad, F.**, Altaleb, M., Abadi, M. 2014. Detection of satellite DNA beta in tomato plants with tomato yellow leaf curl disease in Jordan. Plant Disease 98 (7): 1017.2.

2013

- **Haj Ahmad, F.**, Odeh, W., Anfoka. G. 2013. First Report on the Association of Squash leaf curl virus and Watermelon chlorotic stunt virus with

Tomato Yellow Leaf Curl Disease. Plant Disease
97: 428.

2011 • Anfoka, G., Edwan, H., **Haj Ahmad, F.**, Odeh,
W. 2011. Detection and identification of cereal
viruses in Jordan. Journal of Plant Pathology 93:
2380.

2011 • Al-Musa, A., Anfoka, G., Al-Abdulat, A., Misbeh,
S., **Haj Ahmed, F.**, Otri, I. 2011. Watermelon
chlorotic stunt virus (WmCSV): a serious disease
threatening watermelon production in Jordan.
Virus Genes 43: 79-89.

2010 • Anfoka, G., Edwan, H., **Haj Ahmad, F.** 2010.
First report of barley yellow dwarf virus and
maize dwarf mosaic virus in Jordan. Petria,
Giornale di Patologiadelle Piante 20: 268
Proceedings "13th Congress of the Mediterranean
Phytopathological Union, MPU"; June 20-25,
2010, Rome Italy.

2009 • Anfoka, G., **Haj Ahmad, F.**, Abhary, M.,
Hussein, A. 2009. Detection and molecular
characterization of viruses associated with tomato
yellow leaf curl disease in cucurbit crops in
Jordan. Plant Pathology 58, 754-762.

2008 • Anfoka, G., Abhary, M., **Haj Ahmad, F.**, Rezk,
A., Akad F., Abou-Jawdah Y., Lapidot M.,
Vidavski F., Nakhla M. K., Sobh H., Atamian H.,
Cohen L., Sobol I., Mazyad H., Maxwell, D. P.,
Czosnek, H. 2008. Survey of tomato yellow leaf
curl disease-associated viruses in the Eastern
Mediterranean Basin. Journal of Plant Pathology
90: 311-320.

2008 • Al-Musa A., Anfoka G., Misbeh S., Abhary M.,
Haj Ahmad, F. 2008. Detection and molecular
characterization of Squash leaf curl virus (SLCV)

165

in Jordan. Journal of Phytopathology 156: 311-316.

Conferences

2/2018	• **31st Conference of Molecular Biology of Plants**, (Dabringhausen/Germany) Poster: "Phosphoproteomics Analysis of Systemin Signaling" F. Haj Ahmad, W. Schulze, A. Stintzi, A. Schaller
01/2013	• **2nd trilateral Hohenheim-Middle East Symposium on TYLCV Resistance in Tomato,** (Larnaca/Cyprus) Presentation: "The Molecular Basis of Resistance to *Tomato yellow leaf curl virus* in Tomato"
11/2008	• **Middle East Agricultural Research and Development (MARD) symposium entitled Frontiers in Agriculture: Abiotic and Biotic Stress** (Amman/Jordan) Poster: "The Effect of *Cucumber mosaic virus*, *Potato virus Y* and *Tobacco mosaic virus* Infection on the Gene Silencing Triggered against *Tomato yellow leaf curl virus*" F. Haj Ahmad, G. Anfoka

Stuttgart, 28.04.2019   Fatima Haj Ahmad